GUIDE TO RAPID REVISION

EIGHTH EDITION

Daniel D. Pearlman
Paula R. Pearlman

Eighth Edition Prepared by
Daniel D. Pearlman
University of Rhode Island

and

Edward Steven Shear
University of Rhode Island

New York San Francisco Boston
London Toronto Sydney Tokyo Singapore Madrid
Mexico City Munich Paris Cape Town Hong Kong Montreal

Senior Vice-President and Publisher: *Joseph Opiela*
Marketing Manager: *Christopher Bennem*
Production Coordinator: *Shafiena Ghani*
Senior Cover Design Manager: *Nancy Danahy*
Cover Designer: *Nancy Sacks.*
Electronic Production Specialist: *Jeff Streber*
Manufacturing Buyer: *Roy Pickering*
Printer and Binder: *R. R. Donnelley & Sons—Crawfordsville*
Cover Printer: *Coral Graphics, Inc.*

Library of Congress Cataloging-in-Publication Data

Pearlman, Daniel D.
Guide to rapid revision / Daniel D. Pearlman, Paula R. Pearlman.
— 8th ed. / prepared by Daniel D. Pearlman and Edward Steven Shear.
p. cm.
Includes index.
ISBN 0-321-10757-8
1. English language—Errors of usage. 2. English language
—Rhetoric. 3. Editing. I. Pearlman, Paula R. II. Shear, Edward
Steven. III. Title.
PE1460.P37 2002
808′.042—dc21 2002025475

Please visit our website at http://www.ablongman.com

ISBN 0-321-10757-8

678910—DOC— 080706

CONTENTS

Arranged alphabetically by
Correction Symbols

PROOFREADING SYMBOLS

⊂ CLOSE UP SPACE (Symbol used within the line: wo rd).

INSERT A SPACE (Combined with symbol "∧" used within the line: thetime is).

∧ INSERT MISSING MATERIAL (Symbol used within the line: th t).

ℓ DELETE

¶ PARAGRAPH

⊙ PERIOD

~ or *tr* TRANSPOSE (Symbol used within the line: the to boy).

? UNCLEAR WRITING OR QUESTIONABLE IDEA (when it does not simply refer to use of the *question mark*, for which see p. 82).

PREFACE

The eighth edition of *Guide to Rapid Revision* marks the thirty-eighth year of the text's use in classes of every description, including freshman and advanced composition, technical and business writing, ESL and developmental writing, creative writing; literature, and courses in every discipline that embody our continuing national commitment to improve "writing across the curriculum." Short of a student–instructor hotline, the *Guide* acts as the closest thing possible to an instructor-over-the-shoulder for any student working independently at revision. The new edition has been revised throughout, both for substance and for style. The overall emphases, however, remain the same: brevity, accessibility, and practicality.

A Review of the Major Features

1. *Instant access to information:* An outstanding feature of the *Guide* has always been the alphabetical arrangement of its contents in accord with the correction symbols in common use throughout the country.

Massive cross-referencing reveals the network of logical relationships among the "info-bites" that appear under separate headings. Other major aids to information retrieval are an exhaustive *Index* at the back of the book and the list of *Topical Contents* in the front that facilitates classroom study of whole topics like "Writing Style" and "Sentence Correctness."

2. *Compactness:* Even the longest sections, such as "Commas" or "Variety in Sentence Patterns," take up only a few *pages* as compared to long *chapters* in the usual college English handbook. For the student, such brevity means an effective job of revision in a minimal amount of time.

3. *Focus on problem solving:* There is not a word here that does not contribute directly to revision! Rules and principles of grammar and rhetoric are stripped down to a functional minimum. In the belief that a well-chosen example speaks louder than pages of abstract explanation, the pedagogical emphasis is on the illuminating example, and whatever explanations are

necessary are linked to concrete examples modeling the solution of specific problems.

4. *Clarity:* The style throughout is direct and to the point, neither patronizing nor condescending. The author assumes that the reader is an intelligent seeker of guidance through the brambles of written usage.

New to the Eighth Edition

For the many courses in which research papers are required, the newly revised and expanded sections on documentation (**doc**) provide all the information that a student is likely to need regarding proper formats for the *in-text citation* of outside sources and for the construction of a proper *bibliography.* The most common documentation format for the humanities (MLA), and that most common for the social sciences (APA), are each given extensive treatment in the updated Appendix on Documentation Styles, where students will find examples showing them precisely how to format all the types of information sources they are likely to encounter in their research.

The documentation of those increasingly valuable *electronic* information sources, such as email and the Internet, are here accorded the elaborate treatment they deserve. Instructors who use the *Guide to Rapid Revision, Eighth Edition,* may now, with confidence, dispense entirely with the requirement that their students purchase a separate handbook detailing procedures for documenting information.

I am happy to express my collegial gratitude to Edward Steven Shear, who prepared these new sections on documentation and whose expertise in information science has fitted him ideally to the task.

I am further indebted to Professor Shear for preparing an entirely new introductory section called "An Introduction to Writing as a Process." We hope that this new section, with its emphasis on techniques to help students focus on subject, audience, and aim, and with the techniques it suggests for *prewriting* (helpful methods for achieving that final draft!), will be found useful for both student and teacher. Though revision may always be needed, the thoughtful activity of pre-writing will considerably *reduce* the labor of revision students will need to devote to their final drafts.

I remain deeply grateful for the countless fruitful suggestions offered by instructors and students who have used this text during its past six incarnations, including the reviewers of this edition: J. Robert Baker, Fairmont State College; Valerie Balester, Texas A & M University; Kristine R. Dassinger, Genesee Community College; Michel de Benedictis, Miami-Dade Community College; Patrick McMahon, Tallahassee Community College; William Provost, University of Georgia. I welcome all unsolicited comments concerning the improvement of the Guide. My Internet address is DPEARL@URI.EDU. This edition, too, embodies every helpful criticism I received that did not necessitate major changes in the book's scope and purpose.

D. D. P.

TO THE INSTRUCTOR

I have long felt that the process of revision, central in the development of writing skills, has not been given the full attention it merits among books published for courses in English composition. Most composition handbooks follow a sequential and topical plan designed for *study* of the problems of writing but not necessarily for their immediate solution. Students who turn to these texts for help when following the instructor's suggestions for revision find that they lose time hunting for the passages relevant to their particular problem. In fact, they must often read as much as a full chapter for each error they have made.

The present *Guide*, planned entirely with the realities of revision in mind, gives students *immediate* answers to specific problems, offers sufficient information to solve them, and yet does so with *brevity*. Using the book independently, students may feel as if an instructor were personally going over their papers with them point by point in conference.

Among the major time-saving features of this *Guide* are its compactness and the alphabetical arrangement of its contents. Another convenience is a table of correction symbols that doubles as a table of contents. The table presents the most common symbols used by English instructors throughout the country, and on the inside back cover there is space for students to list extra symbols their instructors may use from time to time. In the text are many realistic examples, often culled from actual student papers, of various types of writing deficiency. Many of these examples are followed by brief explanations that show exactly how to apply the general rule for revision.

Because the *Guide* is designed for independent use by students, you could have your classes spend a period now and then revising their papers, *Guide* in hand, under your direct supervision. The clear, compact treatment of each topic in this *Guide* should enable students to overcome a number of their weaknesses in short order; meanwhile, demands on you for individual help will be reduced to a workable minimum.

Guide to Rapid Revision is valuable both in courses where a traditional handbook of composition is assigned and in those, also, where no such handbook is used. The *Guide* substitutes for the larger handbooks because it contains an adequate treatment, despite its small physical compass, of the basics of English style, usage, and mechanics. It also includes elaborate sections on Documentation, obviating any need for a separate handbook on citation formats for the two most widely used styles (MLA and APA). Although our focus remains revision, the new section titled "An Introduction to Writing as a Process" should help to reduce time spent on revision if students apply the techniques offered there for planning and pre-writing.

A topically organized listing of all sections of the *Guide* is provided on the next page to facilitate your using it as a classroom teaching text. For example, instructors interested in focusing for part of the semester on topics such as sentence correctness or punctuation will find listed under each topic the relevant sections in the *Guide*. It is hoped that instructors will be pleasantly surprised to note the considerable attention paid in so brief a guide to many matters involving writing style, and not only matters of basic correctness.

TOPICAL CONTENTS

Instructors who would benefit from a handbook-style *topical* approach to the concepts covered in this guide will find helpful the following logical groupings (with some inevitable overlap) of related sections:

Writing Style

Sentence Correctness

Punctuation

Points of Grammar and Mechanics

TO THE STUDENT

This book is designed to save you many hours in revising your compositions. Its explanations of English usage are brief, clear, and to the point, and it includes realistic examples that you can use to correct your specific shortcomings. Years of teaching experience have convinced me that most of your writing problems can be eliminated in short order. In keeping the book short, I have tried to include all information that could cast real light on your writing difficulties. Each time you revise, you learn the principles of English usage so that you do not repeat the same mistakes.

HOW TO USE THIS GUIDE

If your instructor uses correction symbols and you are not certain of their meaning, the alphabetically arranged table of correction symbols found at the beginning of this book will tell you what the symbols mean and what page to turn to for help. This book avoids lengthy grammatical analyses of your writing problems. Specific examples—combined with short, concrete explanations—show you how to overcome your weak points. Extensive cross-referencing, such as the advice in the "Paragraph" section to "see **Coherence** and **Transitions**," enables you to find further information related to some special aspect of the problem at hand. Pursue such cross-references if you have time. Assume, however, that the *essential* information you need is already provided in the section you are reading.

For your research needs, the sections on documentation should answer virtually all your questions about how to format the citations and bibiliography needed to show your sources of information.

Special spelling problems are handled in the sections on abbreviations and numbers and in the "Words Often Misused: A Glossary" section under *Diction*. Otherwise, for the usual misspelled word, refer to your dictionary. (Computer spellcheckers are fine, but if your misspelled word happens to be another word—*your* for *you're*, or *it's* for *its*, for example—the spellchecker will not pick up the misspelling.)

You will find this *Guide* a valuable reference for your formal writing and revision needs. *Formal* writing is more conservative in grammar and phrasing than *informal* writing. The emphasis in formal writing is on the objective, impersonal communication of ideas and information, whereas informal writing tends to focus on self-expression—the communication of feelings—and highlights the writer's own personality. Almost all the writing you will be asked to do in school or in the business and professional world is of the formal sort.

Note, however, that written communications vary in *degrees* of formality and informality depending on the audiences for which they are intended. The formal style of articles in scholarly journals, for example, is much more conservative than the style of the average article in *New Yorker* magazine. You rarely see contractions used in scholarly journals, whereas they are quite common in even the most intellectually serious pieces in magazines intended for wide audiences. Your guide in making stylistic choices will always be your own reading experiences with materials that have been written for the kind of audience you wish to reach.

INTRODUCTION TO WRITING AS A PROCESS

This section discusses the *process* of writing and the place of revision within it. The title of this book, *Guide to Rapid Revision*, indicates that its primary purpose is to guide writers in revising and improving their work. Implied is that the writer already knows the stages involved in producing an effective draft. Often, individuals consider revision to be simply the task of correction: fixing "errors," making final improvements—proofreading, in other words.

But experienced writers understand revision as "re-seeing," an ongoing process of reexamination of the paper, including its basic ideas, its overall organization, its sentence structures and phrasing, right down to the hunt for "errors" known as proofreading. The goal of revision is to ensure that you have met all your writing goals as effectively as possible.

WRITING AS A PROCESS

Before you begin to write, you go through a number of stages that enable you to accomplish a first draft. Your first order of business is to think of your writing as a *process* beginning well before you start tapping at your keyboard and ending with a final draft that has survived your *repeated* review. Revision is a part of this process and not merely a matter of cosmetics. The first step of the process is both obvious and often unnoticed: deciding to write in the first place.

The stages in the total writing process may be described as follows:

1. The stimulus
The stimulus is the event that motivates you to write. This may be as simple as a feeling that you want to write a poem, or to respond in writing to an event or issue, or to communicate information, or to persuade readers of your point of view on some controversial topic. In some cases, particularly for students, the stimulus may simply be an assignment.

2. Deciding on Your Purpose

Once a stimulus has been established, the most effective step to take next is to consider the question: "What is my purpose or goal in writing?" Knowing what you wish to accomplish can help determine the content, structure, form, and style of your writing. For example, if you want to inform your readers about a topic they would find unfamiliar, then you must present the relevant information in a clear, logical sequence. You must also be clear about what information is most *important* to get across. Next, if you want to persuade a reader to agree with your position on some current, hotly debated topic, you ought to take into account what objections to your position have already been raised so that you can address them.

3. Identifying Your Audience

In addition to knowing your purpose, know your audience. You would write differently to a lover than to an employer. Different audiences and different contexts require different styles of writing. Start by asking questions such as:

"Who are my potential readers?"
"What do my readers know already?"
"What do my readers need to know?"
"What can I expect my readers to understand and what needs to be explained?"
"How can I most effectively present my ideas to this group?"

Such questioning can maximize the effectiveness of your first draft since you have considered the ability of your audience to receive your message.

Student writers are commonly unsure of their audience. Is your audience only your instructor? Probably not entirely! Few instructors see a writing assignment as merely an exercise to give grades. Instead, writing assignments are usually designed to help you prepare to write in the field of the course. Depending on the type of course and the given assignment, your potential audience may be your classmates, an outside group, professionals in your field, or a variety of other audiences. If you are unsure about the audience to whom you are writing for a class assignment, ask your instructor.

4. Pre-Writing
Pre-writing is the activity you engage in before beginning a first draft. The goal of pre-writing is to clarify your purpose. The most familiar pre-writing activity is outlining. Another common activity is research. See *Techniques of Pre-Writing* (below) for several methods you can use to prepare for writing.

5. Drafting
The more complex or lengthy a writing project, the more drafting becomes a necessity. Do not assume that after you write a draft, revision is only an error-correction session. You may well have to write more than one draft. Creating a first draft is simply the logical consequence of the first four stages discussed: reacting to the stimulus, deciding on your purpose, identifying your audience, and pre-writing.

6. Revision (part 1)
This book provides guidance on improving your writing at the revision stage. However, before you jump to the microscopic process of merely correcting mistakes in spelling, mechanics, grammar, and sentence structure, stand back and look over what you have already accomplished. Have you met your initial goals? First, have you satisfied your motivating purpose? Second, have you presented your ideas in a way that will be effective with your audience? Third, will it be useful to engage in some additional pre-writing activities? For example, you may want to outline what you have written—even if you did not do an outline before drafting. An outline will provide you with a clear summary that may permit you to detect problems in organization or logical coherence, or inadequacies in the presentation of information or supporting evidence. Also, in the process of drafting, you may find that you need additional research. You should focus on these *large-scale* issues as your first step in revision.

7. Revision (part 2) and Editing
After completing your large-scale revisions (sometimes more than once), you are ready to work on the paragraph and sentence level. You will have taken the "top-down" approach, meaning that you have thought about large issues such as purpose, audience, and order of ideas before focusing on sentence-

level issues. This approach is based on efficiency. If you spend a lot of time perfecting a paragraph, but then decide to rewrite or remove it, you have wasted your time. The goal is to deal with the big picture first, then address the details.

TECHNIQUES OF PRE-WRITING

Generally, pre-writing is the activity of preparing to write a draft. This section suggests activities that may help you begin to write—to clarify your purpose, to identify your ideas, to organize your ideas. Other techniques exist, but many writers use some or all of the following.

Freewriting

The idea behind freewriting is to get started writing without worrying about introductions, order, or grammar. Freewriting not only helps you get started, but also helps you identify your ideas at the earliest stage. Often, freewriting leads to drafting.

To start, get a blank sheet of paper and a pen. If you are a computer user, open up a new word-processing document. Next, set a time period to write—even one or two minutes would be good enough at first. Now, start writing about your topic. Do not worry about anything except getting words on the page or screen. Ignore details like spelling and grammar. Jot down words, fragments, ideas, anything related to your topic. If you change your mind about something, don't revise. Instead, write "I've changed my mind . . ." followed by the new idea. *The key is not to stop writing.* If you do not know what to write, just write: "I don't know what to write," or something similar. The goal is just to keep *writing* something. If you get an idea going, stay with it. If the idea dries up, jump to a new one. Stop when you either run out of ideas or your time period is up, whichever comes *last.*

Once you have finished, evaluate what you have. If you did this on a computer, print it out. Next, circle words, phrases, or ideas that you produced. Don't worry if no coherent pattern yet appears. Instead, look for ideas that may help you to form a *general picture*. Make a list of the useful items.

Writing from Prompts

Prompts are questions or statements, addressed to yourself, that you can use to come up with topics worth writing about or to develop topics you are interested in. The goal is to generate ideas. After producing the question or statement, list every answer that comes to mind.

Sample prompts for identifying a topic:

> When I talk to friends, we often talk about ...
> I often find myself trying to convince others that ...
> In class I read about [blank] and wanted to know more.

Sample prompts for developing a topic:

> When I discuss [topic], I am often given the following responses: ...
> When I discuss [topic], I am often misunderstood on the issue of ...
> When I discuss [topic], I want to convince others that ...

The goal is to pursue the prompt that will help you move on.

Brainstorming and Listing

With brainstorming you list every idea you have about your topic no matter how far-fetched. Just get the ideas on paper. Then look back and evaluate what you have. Often, ideas that may seem unimportant may turn out to be major.

Discussion and Rehearsing

Discussing your ideas with others is a useful form of pre-writing, for it can help you clarify your ideas. Another benefit is that someone else may raise questions or issues that you had not previously considered.

Rehearsing is a more personal activity. Run through your mind how you might phrase, explain, or develop part of your paper. This is a form of "mental draft." When you discuss ideas, you are doing the same thing. But in rehearsing, you are not under pressure to respond. You can sit back and tranquilly review your various options.

Outlining

For a pre-writing outline, you will probably have a good idea about your topic. In the revision stage, outlining can help you see the organization of your ideas. Assume, for example, that the topic you have chosen is a description of three main types of professors. Your outline might have the following form:

I. Statement of main idea [There are three main types of professors: slavedrivers, ex-hippies, and humans.]

II. Subtopic I [Slavedrivers]

A. First supporting idea [They assign excess work]

1. Detail 1 [anecdote or example]

2. Detail 2 [workaholic personalities]

B. Second supporting idea [They give low grades]

1. Detail 1 [interested only in top students]

2. Detail 2, 3, etc.

III. Subtopic II [Ex-hippies]

A.

B.

IV. Subtopic III [Humans]

A.

B., etc.

To create an outline, break down your general subject into its main sections, its subtopics or subdivisions (using these headings: I, II, III, etc). Then, under each subtopic, list the main points you want to cover (A, B, C, etc.). If these main points suggest further supporting details, break down your A or B into smaller subsections (1, 2, 3, etc.). *The idea is (1) to be certain you cover all the important material, and (2) to present your material in some reasonably logical sequence.*

Reading and Researching

Often, you have a general idea but need more information before your goal is clear. This is the time for reading and researching a topic. Go to the library. Find general sources about your topic. Use these resources to help refine your idea. You need to find your focus. Once you have a clear focus, write a draft before seeking more information. Then you can restrict your remaining research to what is relevant to your topic.

Idea Maps

To draw an idea map, start with a blank sheet of paper. At the center, jot down your general topic. Then branch out with lines to identify all aspects of this topic. Here is a partial idea map of this very section on "pre-writing":

You would then expand out from the center circle, so that, branching out from "Writing from prompts," you might jot down "identifying topic," "developing topic." From each circled

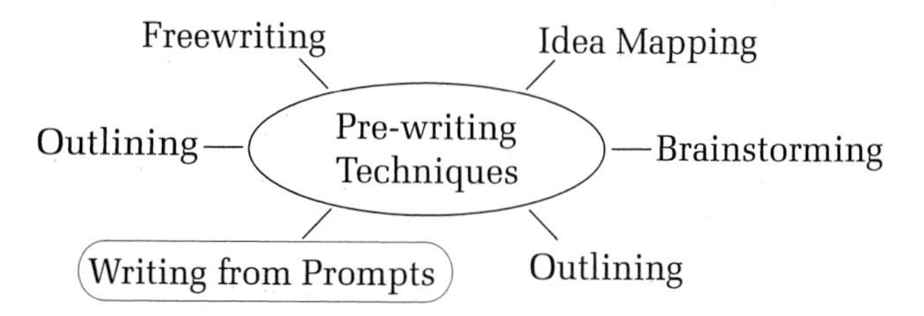

topic you could draw lines mapping further subtopics, depending on how much detail you wanted.

Conclusion

These are only some of the pre-writing techniques that writers use. Perhaps you will find others. In any case, why not eventually try them all to find out what helps you most? If writing is a process, it is also a craft. As with any craft, it helps to have a full box of tools.

ABBREVIATIONS ab

Spell out words in full and do not use telegraphic prose.

As a general rule, do not use abbreviations in formal writing. Some common abbreviations to avoid in your writing are *&, gov't., U.S., U.S.A., thru, OK.* Use *and, government, United States, United States of America, through, okay.*

NOTE: *U.S.* may be used as an adjective (*U.S. foreign policy*), but neither *U.S.* nor *U.S.A.* should be used in formal writing as a noun (*made in the U.S., born in the U.S.A.*).

Especially avoid using *etc.,* short for *et cetera,* meaning *and so forth.* If you really want to say *and so forth,* write it out, but usually you will do better to write out the specific ideas you have in mind rather than to ask your readers to guess at what you mean.

EXCEPTIONS: With proper names, abbreviated titles are preferred: *Dr., Mr., Messrs.* (plural of *Mr.*), *Ms., Mrs., Mmes.* (plural of *Mrs.*), *Jr., Sr., St.* (Saint). The names of academic degrees are also usually abbreviated: B.S. (Bachelor of Science), M.A. (Master of Arts), Ph.D. (Doctor of Philosophy).

Standard abbreviations such as A.M., P.M. (or a.m., p.m.), A.D., B.C.E. and those of certain well-known commodities, organizations, and government agencies such as *TV, VCR, EPA, FBI, NATO,* and *NASA* are also acceptable. When citing a less well-known organization, give the full name of the organization at first mention; then use its abbreviation thereafter, preferably using periods: National Association of Manufacturers, or N.A.M.

NOTE: Most abbreviations are capitalized. When in doubt, refer to a good college-level dictionary.

OVERUSE OF SLASH AND PARENTHESES: Avoid the telegraphic style that results from the use of the slash [/] or parentheses [()] to connect closely related items by writers too impatient to express their ideas in a clear, logical sequence.
ABUSE OF SLASH: Our present administrative policy/program is now bankrupt.
ABUSE OF PARENTHESES: Our present administrative policy (program) is now bankrupt. (See **Parentheses.**)
BETTER: Our present administrative policy *and resulting* program are now bankrupt. [Usually the telegraphic method leaves out the important logical connection between terms that are so hastily joined together.]

NOTE: For *and/or* see "Words Often Misused: A Glossary" under **Diction.** Also avoid the awkward *he/she, she/he* and unpronounceable *s/he* constructions. On avoiding pronouns altogether, see **Sexist Expression.**

ABSTRACT EXPRESSIONS —— abst

1. Add a word or phrase to the abstract term to make it more specific.
2. Replace the abstract term with a word or explanatory passage that is more specific. (See *Vagueness.*)

1. ABSTRACT TERMS

Abstract words and phrases, like *beauty, evil,* and *progress,* have meanings that are somewhat different for each reader. Perhaps you are certain of what you mean by *progress* in a statement such as this: "America has made great progress in the last fifty years." But your reader does not know what you mean until you use a more *specific* expression such as *technological progress* or *moral progress.*

To avoid vagueness, you would no doubt need to explain an expression like *moral progress* even further. Do you mean that there are fewer murders? Do you mean that young people are more moral?

The effect of explaining yourself further is to get more and more *specific* and *concrete* in presenting your ideas. The more precisely you define your ideas, the less you risk being misunderstood.

2. CONCRETE TERMS

A *concrete* word refers to an actual object whose nature is generally known. For example, we all know what *tree* stands for. However, when the tree you are writing about plays an important role in your composition, it is better to use an even more specific concrete expression, like *elm* or *oak.* It is like zooming in for a close-up in movie-making.

ABSTRACT: Politics unfairly determined the results of the recent election. [Many of us, in attempting to explain the ills of society, take the intellectually lazy way out by shrugging our shoulders and blaming everything on politics.]
REVISION: A last-minute public smear campaign together with private blackmail unfairly determined the results of the recent election. [It turns out that the abstract *politics* was screening some rather interesting concrete realities.]

ABSTRACT: The collapse of the Soviet Union was due to the failure of the Soviet system. [Did the *entire* system fail? Did the whole social fabric fall apart? Or does the writer have some specific aspects in mind, as the following revision indicates?]
REVISION: The collapse of the Soviet Union was due to the failure of the Soviet economic and political system.

adj — ADJECTIVE

1. Change the marked word to an adjective.
2. Change the marked adjective to the proper form. (See *Comparison*.)

1. PREDICATE ADJECTIVES

An adjective modifies (describes) a noun or pronoun. Usually, an adjective occurs right next to the word it modifies: the *delicious* coffee. But sometimes the adjective is separated from the word it modifies by a verb, called a linking verb: The coffee smells *delicious*. The adjective *delicious* modifies the noun *coffee*. Adjectives that come after a linking verb are called *predicate adjectives*.

One common writing error that students make is placing an adverb—instead of an adjective—after a linking verb: The coffee smells *deliciously.* The most common linking verbs are all forms of *to be,* such as *is, are, was,* and the following verbs of the five senses: *sound, smell, look, feel, taste.* Use an adjective after these verbs.

INCORRECT: More muscle definition would look *well* on her.
CORRECT: More muscle definition would look *good* on her.

INCORRECT: My roommate felt *badly* about his recent grades.
CORRECT: My roommate felt *bad* about his recent grades.

Feel (Look) Good versus Feel (Look) Well

Ordinarily, *well* is an adverb. Use *well* as an adjective—after *feel, look,* and so on—only when you mean the opposite of ill. It is no compliment to tell a friend that she looks *well* today unless she has just recovered from an illness. If you simply mean that you admire her clothing or makeup, tell her that she looks *good.*

2. CORRECT FORMS OF ADJECTIVES

1. ***Comparative Degree:*** The comparative degree of an adjective is used when you compare two things. The comparative is usually formed by adding *-er* to adjectives of one syllable (great*er*, small*er*) or by placing the word *more* in front of adjectives of more than one syllable (*more* useful, *more* salable). Exception: Two-syllable adjectives ending in *-y* may also add *-er*: *lazier* or *more lazy, angrier* or *more angry, lovelier* or *more lovely.*

NOTE: Do not form the comparative twice:

INCORRECT: Allen is a far *more better* student than Ted.
CORRECT: Allen is a far *better* student than Ted.

2. ***Superlative Degree:*** The superlative degree of an adjective is used when you compare more than two things. Form the superlative by adding *-est* to the end of a one-syllable adjective (great*est*, proud*est*) and by placing the word *most* in front of adjectives of more than one syllable (*most* beautiful, *most* useful). Exception: For two-syllable adjectives ending in *-y*, you may also add *-est*: for example, *laziest* or *most lazy.*
3. ***Irregular Forms:*** The comparative and superlative forms of some adjectives are irregular. *Good* becomes *better* (comparative) and *best* (superlative): *bad* becomes *worse* and *worst.* The following examples illustrate the point:

NOTE: Do not use the superlative in place of the comparative:

> **INCORRECT:** Mary is the *best* of the two writers.
> **CORRECT:** Mary is the *better* of the two writers. [Use *better,* not *best,* if only two individuals are being compared.]
>
> **FEWER, LESS:** Look these up in "Words Often Misused: A Glossary" under **Diction.**

adv —— ADVERB

Change the marked word to an adverb—usually by adding *-ly*.

Adverbs are words that modify (describe) verbs, adjectives, or other adverbs. Most adverbs, although far from all, are made up of adjectives with *-ly* endings. Adverbs limit the meanings of the words they modify by setting specific conditions such as *how* (*unusually* lucky), *when* (left *immediately*), and *where* (far *ahead*).

Certain verbs like *sing, dance,* and *write* often mislead people into using an adjective where an adverb is needed:

INCORRECT: He sang *beautiful.* [His song may have been beautiful, but we want to describe his action.]
CORRECT: He sang *beautifully.*

NOTE: Not all adverbs end in *-ly.* Some common adverbs have unusual forms—*well, rather, very, late, soon, seldom, often, now, later, today, tomorrow*—and some prepositions double as adverbs: He turned *around,* fell *behind,* jumped *up.* The adverbs *well* and *very* are often incorrectly omitted in favor of adjectives:

INCORRECT: She writes *good.*
CORRECT: She writes *well.* [The adverb *well* specifies *how* she writes.]

INCORRECT: They had a *real* good time.
CORRECT: They had a *very* good time. [In careless or casual speech you often hear *real* misused as an adverb, as in *a real nice day.* A better correction than *very* nice might be a single expressive word like *wonderful.* See **Triteness.]**

AGREEMENT —— agr

1. **Make the verb in this sentence agree in number with its subject. If the verb is in the singular, change it to the plural, and vice versa. (See also *-s Error in -s Endings.*)**
2. **Make the pronoun in this sentence agree in number with its antecedent—the word the pronoun refers to.**

1. SUBJECT–VERB AGREEMENT

In the present tense, all verbs end the same in both the singular and the plural—except for the *third-person singular,* where an *-s* is added. The third-person pronouns are *he, she,* and *it*: He moves; she moves; it moves. Most of the time you will be using words that can be replaced by *he, she,* or *it*: *John* moves; *Barbara* moves; *the cloud* moves. Still, the verb ends in *-s*.

All other pronouns, singular or plural, agree with the verb without the *-s*: I *work*; we *work*; you *work*. Plural words that can be replaced by the pronoun *they* also agree with the verb without the *-s*: The *machines* work.

In simple sentences you can easily see how all third-person singular subjects take or agree with *-s* verbs and how all other subjects take the form without *-s*:

- *He* always *speaks* so carefully.
- Our *refrigerator makes* clanking noises. [*Refrigerator* can be replaced by *it*.]
- *They live* right under a volcano.
- *Carol, Iris,* and *John live* very comfortably. [*Carol, Iris, and John* can be replaced by *they*.]

You are likely to make mistakes in sentences when you are not sure what the subject is or when you do not know whether the subject is third-person singular or plural.

Do not be derailed by words and phrases that come between the subject and verb. Find the *simple* subject (the subject stripped of all its modifying words and phrases). *The simple subject is never part of a prepositional phrase.* But prepositional phrases often follow the subject and might confuse you, as in the following sentences:

> **INCORRECT:** The destruction of the world's rain forests *are* the major ecological disaster of our time.
> **CORRECT:** The destruction of the world's rain forests *is* the major ecological disaster of our time. [The subject is *destruction,* not *forests*; the words *of the world's rain forests* constitute a prepositional phrase. You can find the simple subject if you block off, temporarily, all prepositional phrases in the sentence. See **Variety in Sentence Patterns** for definition of prepositional phrase and list of common prepositions.]

NOTE: For the present tense of the verb *to be*, all third-person singular subjects agree with the verb form *is*. All other subjects agree with *are*—except, of course, for the first-person singular *I am*. (See also **-*s* Error in -*s* Endings.**)

> **INCORRECT:** One of the requirements of membership *are* monthly dues payments.
> **CORRECT:** One of the requirements of membership *is* monthly dues payments. [Notice that two prepositional phrases, *of the requirements* and *of membership*, come between the subject *one* and the verb.]

In some cases, normal sentence order is reversed and the subject *follows* the verb:

> **INCORRECT:** After the cheerleaders *come* the band.
> **CORRECT:** After the cheerleaders *comes* the band. [The *band* comes.]

> **INCORRECT:** There *is* two dogs in the park.
> **CORRECT:** There *are* two dogs in the park.

Be careful of sentences beginning with *there* followed immediately by a verb. *There* will not be the subject. The subject will always follow the verb. In the above example, because *two dogs* is a plural subject, it takes the verb *are*.

When singular subjects are joined by *either . . . or* or *neither . . . nor*, use the singular verb. Remember that singular subjects are ones that can be replaced by *he, she,* or *it*; singular verbs usually end in *-s*:

> **INCORRECT:** Neither the mayor nor the police chief *care to* admit that the town has a drug problem.
> **CORRECT:** Neither the mayor nor the police chief *cares* to admit that the town has a drug problem.

If one of the subjects joined by *either . . . or* or *neither . . . nor* is not singular, then the verb agrees with the nearer subject:

INCORRECT: Neither the captain nor the coaches *tries* very hard to win.
CORRECT: Neither the captain nor the coaches *try* very hard to win. [*Coaches,* a *they* word, is nearer to the verb.]

CORRECT: Neither the coaches nor the captain *try* very hard to win.
CORRECT: Neither the coaches nor the captain *tries* very hard to win. [The third-person singular, *captain,* is nearer to the verb; *captain* takes the singular *-s* form.]

NOTE: Collective nouns—like *group, team, squad, family, crew, committee, couple*—are singular nouns that stand for a collection of individuals. Normally they take a singular verb: "The committee *has* adjourned," "The family *adheres* to its traditions." When, however, the focus is on the actions of the individual members within the group, and not the group in general, the noun takes a plural verb: "The family *are* quarreling with one another," "The young couple *were* exchanging kisses," "The crew *were* glad to get out of their sweaty uniforms." (Better: The crew *members* were glad to get out of their uniforms.)

2. PRONOUN–ANTECEDENT AGREEMENT

When the antecedent—the word that a pronoun refers to—is singular, use a singular pronoun. When the antecedent is plural, use a plural pronoun:

CORRECT: *Sandra* knows *she* is smart. [The pronoun *she* refers to *Sandra. Sandra* is the antecedent, the word that the pronoun *she* refers to.]
CORRECT: A short time after buying her *books,* she somehow lost *them.* [The antecedent of *them* is *books.*]

Notice that in both these examples, the pronoun agrees in number with its antecedent: *She* is singular, as is *Sandra*; *them* is plural, as is *books.*

INCORRECT: Vacations offer us opportunities for self-renewal, but *it is* often wasted when we fail to plan ahead for *it*.
CORRECT: Vacations offer us opportunities for self-renewal, but *they are* often wasted when we fail to plan ahead for *them*. [*Vacations,* a plural, is the antecedent of *they*. The correction not only changes *it* to *they* but also changes the verb of the first *it* from *is* to *are*.]

INCORRECT: I like to read a book now and then just for my own pleasure, especially if *they are* short and topical.
CORRECT: I like to read a book now and then just for my own pleasure, especially if *it is* short and topical.

Writers and speakers face the problem of which pronoun to use when antecedents such as *each, everybody, everyone, anybody, nobody, no one, one, either, neither* are singular. In the past, the solution has been to use the third-person singular masculine pronoun:

- Everyone in the class raised *his* hand.

However, this solution leaves out women. Several ways out of this quandary are briefly illustrated under **Sexist Expres-sion, 1.**

AMBIGUITY —————— amb

Revise the ambiguous passage to make it clearly mean one thing only. Ambiguity means *double* meaning or *vagueness* of meaning:

AMBIGUOUS: This morning our bus was *held up* by a pair of red-jacketed men at a construction site. [Was this a *holdup* in the criminal sense?]
CLEAR: This morning our bus was *stopped* by a pair of red-jacketed men at a construction site.

AMBIGUOUS: John asked Bill if *he* could help *him.* [Who needs help, John or Bill?]
CLEAR 1: John asked Bill *to help him.* [In this version, John needs help.]
CLEAR 2: John asked Bill *if he needed his help.* [Here, John is offering to help Bill.]

AMBIGUOUS: Visitors sometimes leave sessions with the president *feeling frustrated and even a bit alarmed.* [This sentence, quoted from a news story, may leave *us* feeling frustrated too! After all, who is feeling frustrated, the visitors or the president? As it stands, there are two possible answers, each shown in the clearly revised versions that follow:]
CLEAR 1: After leaving sessions with the president, visitors sometimes feel frustrated and even a bit alarmed.
CLEAR 2: After visitors leave sessions with the president, he sometimes feels frustrated and even a bit alarmed.

(See **Misplaced Modifier; Pronoun Reference;** and **Vagueness.**)

ap, apos——————APOSTROPHE

Add a missing apostrophe ('), or remove one you have mistakenly used. The apostrophe has three main uses:

1. It marks the possessive case of nouns.
2. It indicates a contraction.
3. It indicates plurals of letters, abbreviations, and numbers.

1. POSSESSIVE CASE OF NOUNS

For nouns, both singular and plural, that do not end in *s*, form the possessive by adding *'s*: the *bird's* nest; the *children's* party; the *person's* name; *today's* weather.

For plural nouns that end in *s*, add the apostrophe only: the *soldiers'* uniforms (uniforms of the soldiers); the *ladies'* coats (coats for ladies); two *months'* time.

For singular nouns that end in *s*, add *'s*. But if the last *s* would be awkward to pronounce, drop it and add only the apostrophe: the *boss's* daughter (daughter of the boss) but *Rameses'* kingdom, *Moses'* leadership.

NOTE: Do not use an apostrophe in the personal pronouns *its, his, hers, ours, theirs, whose.*

2. CONTRACTIONS

Always use the apostrophe to show the omission of a letter or letters in the contracted form of words: *wasn't* (was not), *I've* (I have), *we'll* (we will), *you're* (you are), *it's* (it is), *don't* (do not).

NOTE: As a general rule, avoid contractions in formal writing.

3. PLURALS OF LETTERS, ABBREVIATIONS, AND NUMBERS

Use the apostrophe for plurals of lowercase letters: *n's, x's, p's,* and *q's.* But for capital letters you can follow either of two styles: *Qs* or *Q's*—unless the *s* alone would be confusing, such as in *As.*

Use the apostrophe for plurals of abbreviations containing periods: *B.A.'s, C.P.A.'s, R.N.'s.* But for abbreviations without periods you have a choice of two styles: *VIPs* or *VIP's, VCRs* or *VCR's.*

NOTE: Except in professional and academic degrees, abbreviations tend to omit periods.

You have a choice of two styles for the plurals of numbers: either *5's, 10's,* the *1900's* or *5s, 10s,* the *1900s.*

NOTE: Whenever you choose a style, use it consistently throughout your composition.

art ——— ARTICLE

Most problems with articles—*a, an, the*—involve either (1) use of the incorrect form of the indefinite article or (2) use of articles with uncountable nouns.

1. Use *An* instead of *A*.

The *indefinite article* has two forms, *a* and *an*. It is used as an adjective before a noun. (The *definite article,* also used as an adjective before a noun, has only one form—*the*.)

A is used before words that start with a consonant sound. *An* is used before words that start with a vowel sound. It is the sound, not the first *letter* of a word, that tells you to use *a* or *an*. For example, you write *an hour* because the *h* is silent and *hour* really begins with a vowel sound. On the other hand, you write *a once-in-a-lifetime chance* because *once* begins with a consonant sound (*w*), and you write *a union* because *union* begins with a consonant sound (*y*).

Abbreviations such as M.A. (*Master of Arts*) or SASE (*self-addressed stamped envelope*—an item editors usually require of contributors to their publications) present a special problem. Does one write, "She earned *an* M.A. or *a* M.A."; "I enclose *an* SASE or *a* SASE"? Although usage is not settled on this point, common sense dictates that either may be correct, depending on how the writer wishes the reader to *pronounce* the abbreviated term.

2. In general, do not use articles before *uncountable nouns.* (But do use articles before *countable nouns.*)

Countable nouns refer to people and things that exist as separate—and therefore *countable*—units: professors, ostriches, coins. Use the appropriate article—*a, an,* or *the*—before such nouns:

> **INCORRECT:** I always take notes when *professor* lectures.
> **CORRECT:** I always take notes when *the professor* lectures.
> **INCORRECT:** Old *dog* cannot learn new tricks.
> **CORRECT:** *An* old *dog* cannot learn new tricks.
> **INCORRECT:** *Wind* blew *bouquet* out of his hand.
> **CORRECT:** *The wind* blew *the bouquet* out of his hand.

Uncountable nouns refer to things that are thought of as wholes, as a mass, not as a countable set of units: for example, *abstractions* such as *love, progress, business*; and *substances* such as *air, water, wood, rice.*

a. Uncountable nouns do not normally have *plurals.*
b. Uncountable nouns do not normally have *articles* in front of them.

There are a number of uncountable nouns in English that do have plurals in other languages and therefore often confuse the non-native speaker of English. Examples: *advice, furniture, information, luggage, money.* (Remember, these nouns are singular and therefore take singular verbs: "Your advice *is* valuable"; "Your luggage *has* arrived.")

> **INCORRECT:** *The love*, not *the self-interest*, enables *the society* to exist.
> **CORRECT:** *Love*, not *self-interest*, enables *society* to exist. [Abstract ideas]

> **INCORRECT:** We had *the chicken* and *the rice* for dinner.
> **CORRECT:** We had *chicken* and *rice* for dinner. [Substances]

> **INCORRECT:** She had *the money* to buy *the furniture* but needed *the advice* on where to purchase it.
> **CORRECT:** She had *money* to buy *furniture* but needed *advice* on where to purchase it. [**Note:** If, in the sentence marked "incorrect," *the money* and *the furniture* are each thought of as a specific, *known* quantity or set of items, the words are not being used in their general sense as un-

countable nouns. In such a case, the use of *the* would be correct. The same is true for *the chicken* and *the rice* in the previous example.]

INCORRECT: He needed *the information* about where to eat well in Manhattan.
CORRECT: He needed *information* about where to eat well in Manhattan.

awk——————AWKWARD

Rethink and rewrite the marked passage.

Awkward is a catchall term. It may refer to one specific problem in your writing or any combination of problems. It may point simply to an error in diction (inexact use of a word) or to a much larger problem, such as the lack of coherence in a series of sentences. A similar catchall term is *sentence structure* (SS), which may point to anything from an obvious structural error to a messy passage requiring rewriting.

Upon analysis, some problems marked *awkward* can be given more specific names such as ambiguity, choppy sentences, mixed construction, faulty parallelism, repetition, wordiness. In using the term *awkward,* your instructor probably expects you to recognize what is wrong at a glance without more technical advice. If you see no problem with the passage marked *awk,* make an appointment to see your instructor as soon as possible. If on rereading the passage you find it unsatisfactory, then use your own judgment in rewriting it. Sometimes a passage contains such a severe combination of problems that your instructor simply marks it *awk* in order not to discourage you by listing them all. In rewriting, put aside what you have originally written and focus on the original *thought* you were trying to express. Rethink your ideas as well as rewrite them. Most such problems disappear if, in revising, you make a sincere effort to concentrate on the original *idea* you were after:

> **AWKWARD:** Due to the number of students in college, they appear to be all equal because everyone is experiences the same things.

There are several things wrong with the above sentence. Basically, the statement lacks logical coherence. In addition, the use of *due to* is an error in diction, and *is experiences* results from simple carelessness. To call this sentence *awkward* rightly points the student back to the drawing board for total rethinking and revision:

> **IMPROVED:** There are so many students in college undergoing the same experiences that in many ways they seem to be copies of one another.

> **AWKWARD:** Being an avid fan of country music and having a boyfriend who is a devoted rock fan provides a look at the two different types of music.

It is hard to say specifically what is wrong here. Is *being . . . and having . . .* a double subject? If so, it is plural and does not agree with the singular verb, *provides.* But *being . . . and . . . having . . .* sounds more like a *dangling modifier* looking for a missing subject. Since the two halves of the sentence do not fit together, the problem may be *mixed construction.* Who knows? *Awkward* is probably the best label for such an undeveloped sentence:

> **IMPROVED:** Being an avid fan of country music and having a boyfriend who is a devoted rock fan, I get a good look at both kinds of music.

BRACKETS———br [/]

Add brackets, or if you have used brackets incorrectly, change to parentheses.

1. Use brackets to set off your own explanatory comments from the body of a text that you are quoting or editing:

   "Every village, every town [in Spain] is the centre of an intense social and political life," says Brenan.

If you were to use parentheses instead of brackets around *in Spain,* your reader would think them to be Brenan's words and not yours.

2. Use brackets to avoid parentheses within parentheses:

   Kafka's most extraordinary work begins with the description of a man suddenly changed into a giant beetle (*The Metamorphosis* [Original German edition: Leipzig, 1915]).

cap —— CAPITALIZATION

Capitalize the word or words indicated, or change them to begin with a small ("lowercase") letter if you have used capitals incorrectly.

1. Capitalize proper names. These are the names of specific persons, places, things, races, institutions, organizations: *Joe Fox,* the *East River, Tilden High School,* the *United Nations, Native American.* (Capitalize *high school* or *college* only if part of a proper name, like *Bryant College,* but not if used in a general sense: *I am graduating from college next May.*) The word *the* beginning names of organizations should not be capitalized: *the* United Nations, *the* Police Athletic League.

   Capitalize *Mom, Dad, Father, Mother* when these words are the names used in referring to or directly addressing specific individuals ("Hello, Dad. How is Mom doing?"). Do *not* capitalize when these words are common nouns used to refer to these people as members of the whole *class* of moms and dads (Her *dad* worked hard for a living).

   Capitalize *East, West, South, North* when these words name specific regions (She took a job in the *East*), but do *not* capitalize when they are used as directions (Walk *east* five blocks).

2. Capitalize the first letter of every word beginning a sentence, including the first word of every quoted sentence: *H*e said proudly, "*E*verything is in order."
3. In titles of books, articles, movies, plays, short stories, and poems, always capitalize the first and every word except articles, short prepositions, and coordinating conjunctions of four or fewer letters: *The Old Man and the Sea; Much Ado About Nothing; Life with Father; The Train from Rhodesia; A Prayer for My Daughter.*

CASE——case

Use the correct case of a pronoun.

Many of the errors you make in *case* are carryovers of informal speech patterns into the formal situation of writing, where a high degree of grammatical accuracy is usually expected.

Case is the form a pronoun takes when performing a certain role in a sentence. Three cases exist in English: the subjective case, the objective case, and the possessive case. (For nouns in the possessive case, see **Apostrophe.**) How do you know which case to use for a particular pronoun? That depends on your ability to recognize the subjects and objects in sentences. You probably have fewest problems with the possessive case, and those are usually spelling problems.

In a simple sentence like "She hired him," we see the typical English sentence pattern: subject (*She*) + verb (*hired*) + direct object (*him*). To use case correctly, use the subjective case in positions occupied by subjects and the objective case in positions occupied by objects. Two other sentence positions occupied by objects are important to note: indirect objects and objects of prepositions. Verbs may have both direct objects and *indirect objects*: "She gave *him* (*her, me, us, . . .*) a job." You can tell when *him* is an indirect object if you can "translate" it to mean *to him* or *for him.* "She gave *him* a job" equals "She gave a job *to him.*" Another position for objects is after prepositions (*to, for, of, by, with,* and so on). When the object of a preposition is a pronoun, it must be in the objective case: They voted for *him* and *me*; ". . . for *whom* the bell tolls."

PRONOUNS AND THEIR CASES

	Subjective	**Objective**	**Possessive**
Personal Pronouns			
First Person	I, we	me, us	my, mine
Second Person	*you*	*you*	your, yours
Third Person	he	him	his
	she	her	her, hers
	it	it	its
	they	them	their, theirs
Relative Pronoun	who	whom	whose
	whoever	whomever	

Sample Sentences Analyzed for Uses of Case

- **I wrote her a letter about him, asking her several important questions.** [I (subj.) *wrote* (vb.) *her* (ind. obj.) a *letter* (dir. obj.) *about* (prep.) *him* (obj. of prep.), asking *her* (ind. obj.) several important *questions* (dir. obj.).]
- **I urged her to send me a reply with an extra copy for him.** [I (subj.) *urged* (vb.) *her* (dir. obj.) *to send* (vb.) *me* (ind. obj.) a *reply* (dir. obj.) *with* (prep.) an extra *copy* (obj. of prep.) *for* (prep.) *him* (obj. of prep.).]

Common Case Problems

1. ***The double subject.*** Do not use the objective case in double subjects:

   INCORRECT: *Him* and Claire rehearsed the duet.

CORRECT: *He* and Claire rehearsed the duet. [The subjective case *he* is correct. The test for the correct case is to drop *and Claire. Him . . . rehearsed* sounds wrong.]

2. ***The double object.*** Do not use the subjective case in double objects:

INCORRECT: Kate telephoned both Suzanne and *he.*
CORRECT: Kate telephoned both Suzanne and *him.* [*Him* is a direct object. The test for the correct case is to drop *both Suzanne and; telephoned . . . he* sounds wrong.]

INCORRECT: Bill gave her and *I* the information.
CORRECT: Bill gave her and *me* the information. [*Me* is an indirect object.]

NOTE: Do not use *myself* as a way to avoid choosing between *I* and *me.* (Wrong: Bill gave her and *myself* the information.) The pronouns *myself, himself, herself, ourselves, themselves, yourself, yourselves* are used either as reflexive pronouns (I hurt myself) or intensive pronouns—to provide emphasis (I'll do it myself. You yourselves are to blame!).

INCORRECT: They returned the album to Myra and I.
CORRECT: They returned the album to Myra and *me.* [*Me* is the object of a preposition.]

3. ***Pronoun + appositive as subject.*** *Use the subjective case for sentences beginning with a pronoun plus an appositive in the subject position:*

INCORRECT: *Us* students are very practical people.
CORRECT: *We* students are very practical people. [*Students,* part of the subject of this sentence, is an *appositive,* a noun that renames or identifies the noun or pronoun before it. If you drop the appositive *students,* you can see that *Us . . . are very practical* sounds wrong.]

4. **Than/as + *pronoun.*** Use the subjective case for comparisons ending with a pronoun intended as a subject:

INCORRECT: Hilary skates better than *me.*
CORRECT: Hilary skates better than *I.* [The sentence would logically continue as "Hilary skates better than I *do*" or "than I *skate.*" The subjective case—*I*—is needed because the pronoun after *than* is the subject of an elliptical—unfinished—clause: *I skate.*]

AMBIGUOUS: Jaime likes football more than *me.* [Does Jaime like football more than *he likes* me, or does he like football more than *I do*? Probably the latter!]
CLEAR: Jaime likes football more than *I do.*

NOTE: If the pronoun after *than* or *as* is intended as the *object* of the omitted verb, then it should be in the objective case:

EXAMPLE: John likes him better than *me.* [Think of the sentence with the full elliptical clause included: "John likes him better than *he likes me.*"]

5. **To be + *subjective case.*** Use the subjective case for any pronoun immediately following any form of the verb *to be* (*am, are, is, was,* and so on):

   INCORRECT: It was *her* who borrowed my new skis.
   CORRECT: It was *she* who borrowed my new skis.

6. **Who (whoever)/whom (whomever).** In choosing between *who* (*whoever*) and *whom* (*whomever*), use *who* if the pronoun you want is the *subject* of its own clause. Use *whom* (*whomever*) if the pronoun you want is an *object* in its own clause:

   EXAMPLE: *Who* spilled coffee on my diskette? [Correct. *Who* is the grammatical subject of this question.]
   EXAMPLE: *Whom* do you agree with? [Correct. If you turn the sentence around, you get "You agree with *whom*?" and you can see that *whom* is the object of the preposition *with.*]

   EXAMPLE: *Whom* the Gods would destroy they first make mad. [Correct. *Whom* is the object of the verb *destroy* in the clause "whom the Gods would destroy."]
   EXAMPLE: She avoided *whoever* upset her. [Correct. You would expect the object of the verb *avoided* to be *whomever.*

It is not. The object of *avoided* is the whole clause *whoever upset her. Whoever* is correct because it acts as the *subject* of its own clause, *whoever upset her.*]

7. **Whose/who's *and* its/it's.** Do not confuse certain forms of the possessive case with contractions. *Whose* and *its* imply possession or ownership:

   EXAMPLE: *Whose* down parka is this?
   EXAMPLE: Take the parrot out of *its* cage.

Who's and *it's* are contractions and are used informally to replace *who is, it is,* and *it has*:

   EXAMPLE: *Who's* (*Who is*) the culprit responsible for this vandalism?
   EXAMPLE: *It's* (*It is*) your last chance.
   EXAMPLE: *It's* (*It has*) been a long day.

8. ***Pronoun + gerund.*** Use the possessive case for a pronoun that occurs immediately before a gerund (an *-ing* word used as a noun):

   EXAMPLE: She did not mind *my* having a second helping. [Correct. Do not write "*me* having."]
   EXAMPLE: We look forward to *your* joining us. [Correct. Do not write "*you* joining."]

NOTE: For nouns in the possessive case, see **Apostrophe.**

CHOPPY SENTENCES — choppy

Revise your series of short, choppy sentences by varying your sentence patterns. Do not simply combine your sentences with *ands* or semicolons. The result would be a series of *longer* choppy sentences known as *stringy* sentences. If you master the simple art of using a variety of sentence types, your style will become much smoother:

CHOPPY: She had a very good coat. It was with her almost everywhere. It was a dark-blue woolen coat with a blue lining. It was full length and conservative looking. At one time it had a belt, but that was later lost. On the sleeves some worn spots were starting to appear. They showed how much she used it. It kept her warm on cold nights, and that was what counted most.
SMOOTH: She had a very good coat *that* she took with her almost everywhere. Conservative looking, it was a full-length, dark-blue woolen *garment* with a blue lining. *In spite of* having lost the belt, she used the coat *so much that* worn spots were starting to appear on the sleeve. *As far as she was concerned,* what counted most was that it kept her warm on cold nights.

ANALYSIS: A choppy paragraph of eight sentences is rewritten to form a much smoother paragraph of four sentences that is slightly shorter (sixty-eight words for the smoother version versus seventy-two for the choppy version) even though two new phrases have been added. The italicized words and phrases in the revised paragraph do not appear in the original. For the most part, they reveal logical relationships between thoughts, relationships that are not clearly seen in the original. To revise choppy writing, do not simply combine short sentences into longer ones but use the resources of phrasing and sentence structure *to reveal logical relationships between thoughts.* The main techniques of revision used above are discussed in detail under **Subordination; Transitions;** and **Variety in Sentence Patterns.**

coh ——————— COHERENCE

Completely rewrite the indicated passage. As it now stands, the material does not make clear, logical sense. The parts are not organized in a logical pattern.

Coherence literally means "holding together." Other words for coherence are *organization, order, arrangement,* and *pattern.* When your phrases, sentences, and ideas hold together,

your writing has coherence. In coherent writing, the train of thought is easy to follow. Connections and relationships between ideas are clear. Major ideas stand out from minor points, and ideas of equal importance receive equal emphasis.

LOGICAL ORDER OF IDEAS: Simply throwing down your ideas as they occur to you does not guarantee coherence. The mind very often leaps ahead of the pen, so give your pen time to catch up and arrange your thoughts in a logical sequence. One idea should lead clearly to the next in an orderly, step-by-step pattern with no missing links.

There should be a reason that your third sentence follows the second, not vice versa.

LINKAGE OF IDEAS: A good writer uses various devices to clarify the relationships *between* ideas. A good writer provides links or transitions that lead the reader from sentence to sentence without confusion. Consider the links—the italicized words—between the following two sentences:

Sky-diving is a fast-growing sport. *In fact,* there is a new sports magazine on the market this month *that devotes the entire issue to sky-diving.*

In fact is a transition between the two sentences, a logical bridge telling the reader that factual evidence is at hand to support the first statement. The clause *that devotes the entire issue to sky-diving* links the two sentences by repeating the main idea of the first sentence, including the key word *sky-diving.* Without these two devices you would have the following *incoherent* piece of writing:

Sky-diving is a fast-growing sport. There is a new sports magazine on the market this month.

There are five main ways to link ideas:

1. Transitional words or phrases
2. Repetition of key words or ideas
3. Parallel structures
4. Pronouns
5. Demonstrative adjectives

1. TRANSITIONAL WORDS OR PHRASES

EXAMPLE: More than a dozen condors have bred successfully in captivity. *As a result,* the chance of their extinction seems remote. [*As a result* is a transitional phrase showing the logical connection between the two sentences. Among commonly used transitions are *for example, however, consequently, first of all, on the other hand.* See **Transitions.**]

2. REPETITION OF KEY WORDS OR IDEAS

EXAMPLE: Most economists agree that inflation is caused primarily by *declining productivity.* It is the *declining level of productivity,* the *shrinking output in goods and services,* not the national debt, that threatens the economic base of our country. [Note the repetition of an idea using both the same key words—*declining* and *productivity*—and the synonym *shrinking* for *declining*, as well as another way of saying "productivity": *output in goods and services*. Repetition is okay, but you want to avoid *awkward* repetition. See **Repetition**.]

3. PARALLEL STRUCTURES

EXAMPLE: The three functions that space arrangements must fulfill at our college learning center include *fostering* our students' sense of community, *accommodating* large classroom groups, and *providing* private and semi-private areas for small groups. [It is easy to locate the "three functions" if the descriptions of them employ the same grammatical structure. In this case, three *-ing* words signal the three functions: *fostering, accommodating*, and *providing*. These *-ing* words are used as nouns, each standing for a previous noun, function. An *-ing* word used as a noun is called a *gerund*.]

4. PRONOUNS

EXAMPLE: Professor Donaldson gave the assignment last week. *He* emphasized that *it* would be due today. [Both *he* and *it* are pronouns that refer to nouns in the previous sentence.]

5. DEMONSTRATIVE ADJECTIVES

EXAMPLE: The jury reached a decision in less than twenty minutes. *This* verdict would affect the defendant for the next twenty years. [Demonstrative adjectives point back to previous ideas. There are four demonstrative adjectives: *this, that, these, those*.]

Passage Lacking Coherence

Some people feel that public schools have the right to ban certain books from their libraries. *Huckleberry Finn* has been banned off and on ever since its publication. *The Catcher in the Rye* is often banned. This is in other people's opinion a violation of freedom of speech.

The Above Passage Revised

(*Note:* Added linking devices are in bold type.)

Some people feel that public schools have the right to ban certain books from their libraries. **For example,** *Huckleberry Finn* has been banned off and on ever since its publication. **Another book** often banned is *The Catcher in the Rye*. **Such a banning of books** is in other people's opinion a violation of freedom of speech. [**For example** is a transitional phrase; **such** is a pronoun; and **banning of books** is a repetition of key words.]

NOTE: A special problem in sentence coherence occurs when a quotation is grammatically unrelated to the rest of the sentence. See **Mixed Construction** [the last example under "Mixed Sentence Parts"].

colon :/ — COLON

Place the colon after an introductory statement to call attention to what follows, such as

1. **An explanation.**
2. **A list of items.**
3. **A long quotation.**

1. COLON BEFORE AN EXPLANATION

An explanatory *word* or *phrase* following a statement may be set off with either a colon or a dash:

ACCEPTABLE: The quality of the food served there may be described in a single word: revolting!
ACCEPTABLE: The quality of the food served there may be described in a single word—revolting!
UNACCEPTABLE: The quality of the food served there may be described in a single word *as:* revolting!

If you use *as,* you do not need the colon. *As*, like the colon, points to the explanatory *revolting. As*, however, does not dramatically stop the flow of the sentence the way the colon or dash does.

Generally, the longer the explanatory passage, the more suitable it is to introduce it with a colon:

APPROPRIATE: Small businesses must be protected through appropriate governmental action: the effective and thorough enforcement of antitrust laws in order to maintain competition and prevent agreements and combinations destructive to business.
INAPPROPRIATE: Small businesses must be protected through appropriate governmental action—the effective and thorough enforcement of antitrust laws in order to maintain competition and prevent agreements and combinations destructive to business. [In a formal style, the colon is *preferred* but not absolutely *required*.]

When a full sentence follows a colon, you may capitalize the first word or not, as you please:

CORRECT: Here is our honest opinion: *We* think you are a crackpot.
CORRECT: Here is our honest opinion: *we* think you are a crackpot.

If a *quoted* sentence follows the colon, you *must* begin the sentence with a capital letter:

CORRECT: The sign was perfectly clear: "*No* smoking is allowed in this section."
INCORRECT: The sign was perfectly clear: "*no* smoking is allowed in this area."

If you are quoting a passage that does *not* begin with a capital letter, do not supply one:

> **CORRECT:** She had several choice descriptions of the speech: "wobbling," "waffling," and "wordy" were the mildest.
> **INCORRECT:** She had several choice descriptions of the speech: "Wobbling," "waffling," and "wordy" were the mildest.

NOTE: If you use the words *the following* or *as follows,* expect to place a colon after them:

> **CORRECT:** When I study I proceed *as follows:* First, I review the underlined passages in my textbook; then I accurately copy all technical words and write brief definitions for them.

NOTE: Do not place a colon after the words *as* or *such as* or the forms of the verb *to be* (*is, are,* and so on).

> **INCORRECT:** Her salads include items *such as:* lettuce, tomatoes, carrots, and lima beans.
> **CORRECT:** Her salads include items such as lettuce, tomatoes, carrots, and lima beans.

> **INCORRECT:** The virtues my parents instilled in me *are:* patience, tolerance, and charity.
> **CORRECT:** The virtues my parents instilled in me are patience, tolerance, and charity.

2. COLON BEFORE A LIST OF ITEMS

Use a colon to introduce a series of items at the end of your sentence:

> **CORRECT:** Be sure to take the following things with you on a long ocean voyage: plenty of books, a deck of cards, a chess set, and a warm blanket.

INCORRECT: On a long ocean voyage be sure to take *along:* plenty of books, a deck of cards, a chess set, and a warm blanket.

There is no natural pause after *along*, as there is after *voyage* in the previous example. Do not use the colon to interrupt the normal flow of the sentence. In other words, a colon must be preceded by a complete main (independent) clause:

CORRECT: On a long ocean voyage be sure to take *along* plenty of books, a deck of cards, a chess set, and a warm blanket.
INCORRECT: The things to take with you on a long ocean voyage *are*: plenty of books, a deck of cards, a chess set, and a warm blanket. [The colon after *are* is not needed and interrupts the flow of the sentence.]
CORRECT: The things to take with you on a long ocean voyage *are* plenty of books, a deck of cards, a chess set, and a warm blanket.
INCORRECT: On a long ocean voyage take along things *such as*: plenty of books, a deck of cards, a chess set, and a warm blanket. [Do not use a colon after *as* or *such as*.]
CORRECT: On a long ocean voyage take along things *such as* plenty of books, a deck of cards, a chess set, and a warm blanket. [No punctuation is needed after *such as*.]

3. COLON BEFORE A LONG QUOTATION

If you are quoting a long passage, especially one that consists of two or more sentences, introduce it with a colon, not with a comma.

In "An Apology for Idlers," Robert Louis Stevenson writes: "There is a sort of dead-alive, hackneyed people about, who are scarcely conscious of living except in the exercise of some conventional occupation. Bring these fellows into the country, or set them aboard ship, and you will see how they pine for their desk or their study."

COMMA — c

1. Insert a comma before a coordinating conjunction that connects two main* clauses. *And, but, nor, for, or, so,* and *yet* are coordinating conjunctions.

2. Insert a comma after sentence parts that come before the main clause, especially long phrases and subordinate clauses.

3. Set off parenthetical (nonrestrictive) sentence parts with commas.

4. Insert commas between words, phrases, and clauses in a series.

5. Use a comma to separate coordinate adjectives.

6. Do not use unnecessary commas.

The following six instructions for using the comma will solve practically all your comma problems:

1. *Use the comma before coordinating conjunctions* (and, but, nor, for, or, so, *and* yet) *that join two main clauses.*

*Main clauses are also called *independent* clauses.

- The instructor encouraged her students to ask questions in class, *and* she noted that those who did were often the ones who did best on her exams.

- He thought he would be snapped up by a major corporation, *but* he finally settled for anything he could get.

  EXCEPTION: If the main clauses are very short, you do not have to separate them with a comma:

- I never had *and* I never will.

2. ***Use the comma after sentence elements that appear before the main clause, such as subordinate clauses and long phrases.***

- *When the instructor spoke to the student,* she asked him whether he studied very much. [The words in italics are a subordinate clause. The term is defined under **Variety in Sentence Patterns,** p. 131.]

- *Shaking his head,* the student replied that his roommates kept the television set blaring day and night. [The words in italics are a participial phrase. The term is defined on p. 133.]

- *As a solution to the problem,* the instructor recommended the temporary removal of a few parts from the set. [The words in italics are a long prepositional phrase. See p. 133.]

  EXCEPTION: Most short prepositional phrases that come before a main clause are not followed by a comma:

- *After a moment* the student admitted that television wasn't his only distraction from studying.

Certain introductory words and phrases, such as *for example, in short, in fact, however,* and *consequently,* are used to form a bridge, or transition, from one sentence to another and are followed by a comma:

- *In short,* the student admitted that he simply disliked the course. (See **Transitions.**)

NOTE: If a subordinate clause *follows* the main clause, normally you do *not* separate them with a comma:

- No fishing boats ventured out that day because the water was too rough. [No comma between *day* and the subordinate clause beginning with *because.*]

But modern usage is flexible. Read your sentence aloud. If you hear a definite *pause* between the main and subordinate clauses, you may separate them with a comma:

- I knew he deserved his punishment, although I admit I did feel a moment of pity for him.

3. ***Use commas to set off parenthetical sentence elements.***

NONRESTRICTIVE ELEMENTS: A sentence element is parenthetical, or *nonrestrictive,* if it supplies information that is not essential to the clear meaning of the sentence. In the following examples, the nonrestrictive elements are italicized:

- Modern automobiles, *which are smaller and more fuel efficient than ever,* strike me as more practical and attractive than the older gas guzzlers.

- She is, *I agree,* a good sport.

- Soil erosion, *the loss of water-storing topsoil,* turns land into desert.

To test whether an element is parenthetical, remove it from the sentence. If the basic idea of the sentence remains the same, then the element you have removed is parenthetical and should be set off by commas. Read the previous examples without the words in italics, and you will find that the ideas of the original sentences remain unchanged.

RESTRICTIVE ELEMENTS: Restrictive sentence elements are necessary to the meaning of the sentence, as in this example:

- Everyone who is hard of hearing should wear a hearing aid.

Notice that the clause *who is hard of hearing* is essential to the meaning of the sentence. If you remove it, the basic idea of the sentence is distorted. Restrictive elements are not set off from the rest of the sentence by commas.

4. ***Use commas between items in a series.*** A series consists of three or more elements, which may be single *words, phrases,* or *clauses* (these last three italicized words are in a series):

- The basement was *dark, damp,* and *cold.*

The formula for the series is *a, b,* and *c.* Also acceptable in formal writing is the formula *a, b* and *c*, where there is no comma between the last two items in the series: The basement was *dark, damp* and *cold.* Whichever form you choose, try to use it throughout a piece of writing. (See **Parallelism.**)

- He stumbled down the stairs, across the room, and through the doorway. [A series of three prepositional phrases.]

- I whistled shrilly, I listened in vain, and I turned sadly away. [Three main clauses, if they are short, may be connected in a series by commas.]

- To ease the housing shortage, we need to know *what measures we can take, how we can fund them,* and *who can get the job done most efficiently.*

5. ***Use commas between coordinate adjectives that come before a noun.*** Coordinate adjectives are adjectives that stand in *equal* relation to the noun they modify:

- She is an *old, faithful* servant.

- Look at his *clear, twinkling* eyes.

The test for coordinate adjectives is to insert the word *and* between them and omit the comma. If the adjectives are coordinate (equal in rank), you will feel no awkwardness: *clear and twinkling* eyes.

The test shows that the following examples are not coordinate adjectives: a *small living* room, a *little old* man. The last adjective in each pair is really treated as part of the noun. It would be awkward to say *a small and living room* or *a little and old man.* Where you can insert the *and,* use the comma. Where you cannot insert the *and,* omit the comma.

NOTE: For the use of commas with quotation marks, see **Quotation Marks.**

6. ***Do not use unnecessary commas.*** If you use *too many* commas, you will probably find your types of errors discussed below.

a. Do not use commas to separate subjects from verbs and verbs from their objects or complements—unless nonrestrictive elements come between them. (For "restrictive" and "nonrestrictive" elements see 3 above.)

> **ERROR: (Comma between subject and verb):** The cat-and-mouse game between clever computer criminals and harassed security experts, threatens never to end. [The simple subject is *game.* Its verb is *threatens.* No comma should separate them. The long phrase *between . . . experts* is restrictive, that is, essential to defining the subject, and therefore should not be set off by a comma on *either* end.]

> **ERROR: (Comma between verb and its complement):** For me, the basic ingredients of a good spaghetti sauce are, fresh garlic cloves and a first-press olive oil. [The comma after the verb *are* is not needed.]

b. Do not use commas to separate restrictive elements from the rest of the sentence. (See 3 above.)

> **ERROR:** Paul Ehrlich's book, *The Population Explosion,* warns humankind against overburdening the "carrying ca-

pacity" of the planet. [The phrase *The Population Explosion* is restrictive, that is, essential to the meaning of the sentence, and should not be set off by commas. Commas here imply that this is Ehrlich's *only* book. If it were, then the book's title would be parenthetical or non-essential information. But Ehrlich has written many books, not just this one. The sentence should therefore read: "Paul Ehrlich's book *The Population Explosion* warns . . ."]

c. Do not use a comma before a parenthesis.

ERROR: We put new locks on our doors, (in spite of the cost) for fear of burglary. [Omit comma after *doors.*]

d. Do not use a comma between adjectives if you cannot smoothly insert the word *and* between those adjectives instead. (See the discussion of "coordinate adjectives" in 5 above.)

ERROR: They discovered an ancient, Egyptian tomb. [You cannot smoothly write "an ancient *and* Egyptian tomb." Therefore you cannot use a comma between these adjectives either.]
ERROR: He was a funny, little fellow. [Omit the comma.]

e. Do not use a comma after the last item in a series. (See 4 above for the correct use of commas *between* items in a series.)

ERROR: Many computer researchers feel that terms such as telepresence, artificial reality, and immersive simulation, are preferable to the more commonly used "virtual reality." [The comma after *immersive simulation,* the last in a list of three items, should be omitted.]
ERROR: Gold, silver, and platinum, tend to have widely varying market values. [Omit the comma after *platinum.*]

f. Do not use a comma before the *and* of a compound predicate.

ERROR: College courses in science fiction are proliferating, and are increasingly being regarded as worthy of serious literary study. [Omit the comma before *and.* If the subject *they* came after *and,* the comma would be correct, according to comma rule 1 above, in which the comma is used "before coordinating conjunctions that join two main clauses." But in this example *and* is connecting only the predicate part of a sentence—the verb-part left when a subject is missing—to the predicate part of the whole main clause in front of it. In other words, *and* is just joining the two parts of a compound predicate, and therefore no comma should go before it.]

ERROR: Unemployment has been widespread, and has not been so severe since the Great Depression of the 1930s. [Omit the comma after *widespread.*]

g. Do not use a comma with a question mark or an exclamation point.

ERROR: "Are you coming to the party?," she asked point-blank. [Omit comma after *party?*]
ERROR: Fresh fruit!, he thought, already smelling the apples. [Omit comma after *fruit!*]

h. Do not use a comma after *and, but, or, nor, for, yet, so.*

ERROR: The house was lovely, but, it was too big for us. [Omit comma after *but.*]
ERROR: I knew it was wrong, yet, I was tempted to do it. [Omit comma after *yet.*]

CS — COMMA SPLICE

Change the comma to a period or to a semicolon. Do not join, or splice, two separate sentences with a comma. The comma splice, a type of run-on sentence, seriously handicaps readers by preventing them from distinguishing between the end of one thought and the beginning of the next:

COMMA SPLICE: New York is a busy industrial city, thousands of cars and trucks move through it every day.
REVISION 1: New York is a busy industrial city. Thousands of cars and trucks move through it every day.

This first revision changes the comma splice to two sentences by changing the comma after *city* to a period and capitalizing *thousands.* You could also correct the error by changing the comma after city to a semicolon:

REVISION 2: New York is a busy industrial *city; thousands* of cars and trucks move through it every day. [Before substituting a semicolon, be certain how to use it. See **Semicolon.**]

NOTE: To avoid splicing two separate sentences with a comma, learn to recognize what makes up a separate *sentence.* At the heart of a sentence are a *subject* and a *verb.* (In the example just given, *New York* is the subject of the first sentence, and *is* is the verb.) To test further for a sentence, say your group of words out loud. If it *sounds like a complete statement* (and has a subject and a verb), it is likely to be a *sentence.*

A second type of comma splice occurs in sentences beginning with words (called *conjunctive adverbs* when they link main clauses) such as *therefore, however, then, nevertheless, moreover, also, still, thus,* or with expressions such as *in fact, for example, that is, on the other hand, in other words.* These are transitional words or phrases that begin a new main clause or a new sentence. Most often the main clause beginning with such an expression should be linked with the previous main clause by a semicolon:

COMMA SPLICE: We packed all our luggage, then we were on our way to the airport.
REVISION: We packed all our luggage; then we were on our way to the airport. [Changing the comma after *luggage* to a semicolon removes the comma splice.]

COMMA SPLICE: He was an excellent computer programmer, however, he frequently failed to show up for work.
REVISION: He was an excellent computer programmer; however, he frequently failed to show up for work.

NOTE: Sometimes a conjunctive adverb is *not* found at the beginning of its clause. In such a case, do *not* set it off with a semicolon:

INCORRECT: He was an excellent computer programmer. *Frequently; however*, he failed to show up for work.
CORRECT: He was an excellent computer programmer. *Frequently, however*, he failed to show up for work. [In such cases, as this example shows, it is also best to separate the two main clauses with a period rather than a semicolon.]

COMMA SPLICE: I have always loved sports, in fact, I was once the youngest member of my team in the Little League. [Place a semicolon after *sports*.]
REVISION: I have always loved sports; in fact, I was once the youngest member of my team in the Little League.

COMPARISON ——— comp

Add the word or words needed to complete the comparison. Incomplete comparisons lead to absurd or illogical statements. (For the irregular comparative forms of *good* and *bad*, see *Adjective*, 2.)

ILLOGICAL: The traffic in New York City is worse than Chicago. [*Traffic* is illogically compared to a city!]
REVISED: The traffic in New York city is worse than *the traffic in* Chicago.
BETTER: The traffic in New York City is worse than *that* in Chicago. [Use a pronoun to avoid awkward repetition.]

ILLOGICAL: In this poem Robert Frost expresses ideas different from most other poets. [*Ideas* are illogically compared to *poets.*]
REVISED: In this poem Robert Frost expresses ideas different from *those of* most other poets.

INCOMPLETE: The auditor's income was as high, if not higher than, the company president's. [The complete phrase should be *as high as.*]
REVISED: The auditor's income was as high *as,* if not higher than, the company's president's.

INCOMPLETE: Nolan Ryan has struck out more batters than *any* pitcher in the major leagues.
REVISED: Nolan Ryan has struck out more batters than *any other* pitcher in the major leagues.

INCOMPLETE: Luis received a higher score than *anyone* in the class.
REVISED: Luis received a higher score than *anyone else* in the class.

dang——DANGLING MODIFIER

1. Change the dangling element into a subordinate clause by adding a subject and verb.
2. Change the main clause so that the subject is correctly modified by the dangling modifier.

The modifier in your sentence *dangles* because it does not clearly and logically relate to another word in the sentence. Use either one of the above changes to revise the sentence:

DANGLING: *When sitting,* my shoulders tend to slouch back. [*I*, the logical subject of the modifier, does not appear in the sentence. As now written, the sentence says that *my shoulders* are sitting.]

REVISION 1: *When I sit,* my shoulders tend to slouch back. [This revision changes the dangling elements into a subordinate clause. *Subordinate clause* is defined on p. 132.]
REVISION 2: When sitting, *I* find that my shoulders tend to slouch back. [This revision makes the subject of the main clause, *I,* agree with the dangling element.]

Note that the introductory phrase should logically modify the noun or pronoun *immediately following the comma.* Further, that noun or pronoun should always be the *subject* of the main clause.]

DANGLING: *To type well,* your legs must be in the correct position. [Are *your legs* doing the typing?]
REVISION 1: *If you want to type well,* your legs must be in the correct position.
REVISION 2: To type well, *you must keep* your legs in the correct position.
REVISION 3: To type well, *keep* your legs in the correct position. [In imperatives—statements giving commands—the subject pronoun *you* is implied although not openly stated.]

DANGLING: *Going home,* it started to drizzle. [Where is the subject who is *going*?]
REVISION 1: *As I was going home,* it started to drizzle.
REVISION 2: Going home, *I felt it starting* to drizzle.

DANGLING: *Fearful of a threatened lawsuit,* his decision to pay me back was wise.
REVISION: Fearful of a threatened lawsuit, *he wisely decided* to pay me back. [*He,* not *his decision,* was fearful.]

DANGLING: *At the age of three,* my mother discovered I had a speech impediment. [Was mother *really* three when she discovered this?]
REVISION: *When I was three,* my mother discovered I had a speech impediment. [In this case, there is simply no

smooth way of revising that keeps the dangling phrase *At the age of three* unchanged.]

dash ——————— DASH

Insert or delete a dash.

For the most part, avoid using the dash if commas or parentheses will serve equally well. Use the dash to mark an abrupt shift in thought, to emphasize a parenthetical element, or to ensure a clear reading. However, using the dash too frequently for emphasis becomes monotonous. (See **Emphasis.**)

NOTE: Most keyboards are equipped with only a hyphen and not a dash. To type the dash, use two strokes of the hyphen key [- -]. Leave no space before or after the dash. (See **Colon, 1.**)

Examples of the proper use of the dash:

- "I would like—no, as a matter of fact, I wouldn't." [Abrupt shift in thought.]

- "I must admit—since you force me to tell you—that my opinion of you is not very high." [Dashes set off a parenthetical element emphatically. Parentheses muffle and make unemphatic the material they enclose.]
- The rise of world-scale crises—ubiquitous political atomization, the shrinkage of nonrenewable natural resources, runaway environmental pollution—is the legacy of this century to the next. [The words from *ubiquitous* to *pollution* are a single parenthetical element and normally would be set off by a comma at either end. But the passage itself contains commas, and you might not immediately see that the three items it contains are *examples* of "world-scale crises" and not three extra items in an equivalent series of four (starting with *crises*). To ensure a clear reading, therefore, dashes are used to signal the parenthetical nature of the three items.]
- Intel Corporation's original Pentium computer chip was discovered to harbor a bug—a tiny programming defect. [A dash may be used to introduce a *brief* explanation, but see **Colon, 1.**]

DICTION ——— d

Change the word or phrase you have used to one that is more exact in *meaning*, to one that is less *wordy*, or to one that is more suited in *tone* to the rest of your essay.

Certain errors in diction (word choice) recur frequently. Check to see if your error is dealt with in "Words Often Misused: A Glossary" that appears in this section. In any case, the following suggestions are a guide for correcting and avoiding mistakes in diction:

1. Check the exact meaning of the word you have used in a large modern dictionary ("College-edition" size, at least). You may find in some cases that your problem is spelling, as, for example, confusing *accept* with *except*. This and other

common spelling errors are treated below in "Words Often Misused: A Glossary."

2. See a sample list of common wordy expressions under **Wordiness.**
3. Sometimes the word you choose does not fit the *tone* of the rest of your essay. Tone is the attitude the writer takes toward the subject. The tone may be solemn, humorous, conciliatory, angry, informal, technical. It may reflect any emotional or intellectual attitude imaginable. In a formal essay, for example, it would not be suitable to use terms like *guy, mom, dad,* or *dude.*

Good dictionaries give a variety of labels to words. Check your dictionary to see whether the word you have used is labeled as slang, regional, technical, informal, or nonstandard. Nonstandard words are out of place in the formal style of standard written English that is generally expected of you. If the particular usage of a word is *standard* (generally acceptable in cultivated speaking and writing), it will not be labeled. Some words may have different labels for different meanings: For example, a word like *cool* could have one definition with a standard meaning and another definition with a slang application.

(See **Slang** and **Jargon.**)

Words Often Misused: A Glossary
(See also *Diction*)

Accept, except. *Accept* means *to receive* or *to agree* to something: "I *accepted* his offer." *Except,* used as a verb, can mean only *to exclude*: "He was *excepted* from the list of prize winners."

Affect, effect. The problem here is almost always a spelling error. The verb is *affect* and it means *to influence*: "Her speech *affected* many people." The noun is effect and means *result*: "The *effect* of the blow was to split the stone in half."

Allusion, illusion. An *allusion* is an indirect reference; an *illusion* is a false or deceptive notion.

Alot, a lot. *Alot* is simply a misspelling of *a lot.* In formal writing, however, it is usually better to use *many, much,* or *very much* instead of *a lot.*

Alright. In formal English, this is an unacceptable spelling of *all right.*

Among, between. Ordinarily, *between* is used when only two items are spoken of: "I divided the food *between* the cat and the dog." *Among* relates to more than two items: "The prize money was divided *among* the three winners."

Amount, number. When things or people can be counted individually, use *number*: "I saw a large *number* of students in the hall." When you are referring to a quantity of something that is not thought of as individual, countable units, use *amount*: "A large *amount* of gold was discovered in the mountain."

And/or. Use this compound sparingly and only in a highly technical or legalistic context.

Anyways, anywheres. Use the standard forms *anyway* and *anywhere.*

Around. Do not use the colloquial *around* in expressions like "He left *around* ten o'clock." "I can recite *around* fifteen poems." Use *about*: "He left *about* ten o'clock." "I can recite *about* fifteen poems."

As. *As* in the sense of *because* is often not as clear as *because, for,* or *since.* "*I* would like to leave *because* [not *as*] I'm tired." (See *Like.*)

At. (See *Where at.*)

Awhile, a while. After a preposition, spell as two words: "I slept for *a while.*" Otherwise spell as one word: "I slept *awhile.*"

Bad, badly. After the verbs *feel, look, taste, smell, sound,* use the adjective *bad,* not the adverb *badly.* (See **Adjectives, 1.**)

Because. (See *Reason is because.*)

Because of the fact that. The phrase *because of the fact that* is unnecessarily wordy. Simply use *because.*

Being as, being that. *Because* or *since* are preferred in standard English.

Beside, besides. *Beside* means *at the side of*; *besides* means *in addition to.* "*Besides* chicken, we ate roast beef and bananas as we sat *beside* the stream."

Bust, busted. These are slang forms of the verb *burst.* Use *burst* in present and past tenses. *Bursted* is nonstandard.

Capital, capitol. Use the *-al* version when you mean a *capital city* (like Providence, Rhode Island), *capital punishment* (the death penalty), or *capital* in the sense of assets. Use the *-ol* version when you mean a state's *capitol*—the building a state legislature conducts its business in—or the *Capitol* in Washington, D.C., where the U.S. Congress meets.

Centers around, centers on. In formal English use *centers on*: "The global economy still *centers on* oil production."

Choose, Chose. A spelling mixup. *Choose* is the present tense of the verb *to choose*; *chose* is the past tense. See also *Loose, lose.*

Compare to, compare with. To *compare to* means to find resemblances in things that are otherwise quite different: "He compared the coffee *to* mud." To *compare with* means to find similarities and differences between two things that are of the same sort: "He compared the female students *with* the male students and found the females brighter."

Complected. Use *complexioned. Complected* is nonstandard for *complexioned.*

Could of. This is a misspelling of the contraction *could've,* which stands for *could have.* [As a general rule, avoid contractions in formal writing.]

Data. This word is a Latin plural (singular, *datum*) and is often used in English with plural verbs and pronouns: "*These* data *are* out of date." Many people accept its use in the singular: "*This* data is no longer useful." (See *Phenomena; Strata.*)

Don't. *Don't* is a contraction of *do not* and should not be confused with *does not* or *doesn't.* Nonstandard: "He don't mind insults."

Standard: "He *doesn't* [or *does not*] mind insults." Bear in mind that contractions are generally unacceptable in formal writing.

Due to. Use *due to* only to connect a noun construction with another noun construction: "*Rickets* [noun] is due to a vitamin D *deficiency* [noun]." Do *not* use it to connect nouns with main clauses: Wrong: "Due to an *accident* [noun] *the traffic was backed up for miles* [main clause]." (If you are uncertain, use *because of* or *caused by,* whichever fits.)

Due to the fact that. Avoid being windy. Use *because.*

Effect. (See *Affect, effect.*)

Enthuse. In formal English, it is better to use *to be enthusiatic.*

Equally as good. Drop the *as* and write *equally good,* or use *just as good.*

Etc. This is short for the Latin *et cetera,* meaning *and so on* or *and so forth.* Avoid *etc.*: It is often a substitute for precise and detailed thinking. (See **Abbreviations.**)

Except. (See *Accept, except.*)

Farther, further. *Farther* is often preferred to express extent in *space,* whereas *further* is preferred to express extent in *time* or *degree*: "We walked *farther* into the woods." "He went *further* in condemning him than anyone expected."

Favorite. Because *favorite* is already a superlative meaning "most liked, most favored," to combine it with another superlative, "*most* favorite," is redundant. See also entry for **Unique.**

Fewer, less. When referring to separate items that can be counted, use *fewer*: "You make *fewer* mistakes now than when you started." *Less* refers to the degree or amount of something we consider as a whole and not as a series of individual items: "I have *less* money now than when I started."

Flaunt, flout. *To flaunt* is to show off, as in the TV commercial of some years back: "If you've got it, *flaunt* it." *To flout* is to show contempt for: "He *flouted* all the rules and did things his own way."

Hadn't ought. *Hadn't ought* is nonstandard for *should not.* Instead of writing, "I *hadn't ought* to have gone," write, "I *should not* have gone."

Healthful, healthy. Whatever *gives* health is *healthful* ("a *healthful* climate"), and whatever *has* health is *healthy* ("a *healthy* person").

Illusion. (See *Allusion, illusion.*)

In regards to. Use *in regard to.*

Irregardless. The proper form is *regardless.*

It's. A contraction of *it is* and a common misspelling of the possessive pronoun *its*: "They examined *its* [not *it's*] contents." Use *it's,* and other contractions, only in *informal* written English and in recording actual *speech.*

Kind of, sort of. *Kind of* and *sort of* are informal expressions. In formal written English, use *somewhat, rather, a little*: "She was *somewhat* [not *kind of*] annoyed."

Lay, lie. When you mean *to put,* use *lay.* The forms of *to lay* are "I *lay* the book down" (present), "I *laid* the book down" (past), and "I *have laid* the book down" (present perfect). When you mean *to recline,* use *lie.* The forms of *to lie* are "I *lie* in my bed" (present), "I *lay* in my bed" (past), and "I *have lain* in my bed" (present perfect).

Layed. Incorrect spelling of *laid.* See *Lay, lie.*

Lead, led. A spelling mixup: The past tense of *to lead* is *led,* not *lead.*

Less. (See *Fewer, less.*)

Like, as, as if. *Like* is a preposition and is properly used in a phrase such as the following: "He looks *like* my father." It is improperly used when followed by a clause.

MISUSE: "It looks *like* my father enjoys your company."

NOTE: A clause is a group of words containing a subject and a verb. In the example just given, *father* is the subject and *enjoys* is the verb of the clause "my father enjoys your company."

REVISION: Change *like* to *as if:* "It looks *as if* my father enjoys your company." In the sentence "I behaved *like* I was told to," change *like* to *as,* "I behaved *as* I was told to."

Loose, lose. A spelling mixup: *Loose* (pronounced *loos)* means "slack," as in "a loose knot." *Lose* (pronounced *looz*) is the verb "to lose," as in "to lose a fortune."

May of, might of. These are misspellings of the contractions *may've* and *might've* that in formal English should be spelled out in full: *may have, might have.*

Media. *Media* is the plural form of *medium*; it takes a plural verb: "Some advertising *media* are morally harmful, and the *medium* that sins the most in this respect is television." A singular verb with *media* is nonstandard. (See *Data; Phenomena; Strata.*)

Mighty. Use a standard word like *very:* "I was *very* [not *mighty*] tired."

Most. Use *almost:* "I saw them *almost* [not *most*] every day."

Must of. This is a misspelling of *must've,* a contraction of *must have.* In formal writing use *must have.* (See *could of, may of.*)

Off of. Drop the *of.*

Phenomena. In formal English, *phenomena* is the plural, *phenomenon* the singular.

Principal, Principle. *Principal* can be used as an adjective (*chief, main, highest-ranking*: She has the *principal* role in the play) or noun (referring to a *leader*, as of a high school; or referring to *capital*, in financial usage). It has nothing to do with the noun *principle*, meaning a basic truth or belief (the principles of democracy), or a scientific rule or law (the principles of biomechanics).

Quite. Do not overuse *quite* to mean *very,* as in *quite good, quite hard.*

Real. Keep expressions like *real good* and *real exciting* out of your written English. Use *really good* and *very good.*

Reason is because. In informal usage you may hear: "The *reason* I told you *is because* I can trust you." *Because* is redundant. See also the example under **Mixed Construction.** For formal writing, revise as follows. Method 1: The reason I told you is *that* I can trust you. [*Because* changed to *that.*] Method 2: I told you because I can trust you. [The sentence has been recast.]

Set, Sit. Do not use *set* (to *place* something somewhere: "I *set* my suitcase down") when you mean *sit* or *sat* ("She was glad to *sit* down"). Nonstandard: "Johnny stormed in and *set* down at the bar" Standard: "Johnny stormed in and *sat* down at the bar."

Should of. This is a misspelling of *should've,* a contraction of *should have.* (See *could of.*)

So. (1) Do not overuse *so* as a conjunction joining main clauses. (See **Subordination.**) (2) Do not use *so* where you could use *so that.* Change "I came to visit you *so* we could have a chat" to "I came to visit you *so that* we could have a chat." (3) Do not overuse *so* as an intensifier: "I was *so* disappointed." "She is *so* nice, isn't she?" Try substituting *very* or *extremely.*

Sort of. (See *Kind of, sort of.*)

Strata. Use *strata* only as a plural, not as a singular, noun. The singular is stratum. Nonstandard: He came from an extremely disadvantaged *strata* of society. Standard: He came from an extremely disadvantaged *stratum* of society.

Sure. Use *certainly* or *surely*: *"I certainly* [not *sure*] was tired."

Their, There, They're. Do not confuse the possessive pronoun *their* (*my, your, their*) with the *there* that points to a particular place (*She lives there*), or with the idiomatic expression *there is, there are* (*There are many varieties of English*), or with *they're,* a contraction of *they are.*

Then, than. *Then* is sometimes a misspelling of *than. Than* is used in comparisons: "They would rather die *than* surrender." *Then* means *consequently* or *as a result,* or it refers to time and means *next* or *at that* time: "If the sun is a dying star, *then* the Earth is doomed to extinction." "He came, he saw, and *then* he turned around and left."

To, Too, Two. *To* is a preposition. See **Variety in Sentence Patterns** ("Using a Prepositional Phrase"). *Too* means *also.* Do not confuse the *two* (the number 2).

Try and. *Try and* is an informal version of *try to*: "I am going to *try and* help my neighbor." In formal English write: "I am going to *try to* help my neighbor."

Unique. Because *unique*, meaning "unlike any other," describes the quality of being beyond comparison, do not combine it with adverbs of comparison or degree like "more unique," "most unique," and "very *or* rather unique." It makes no sense to write "Jorge's style is *more unique* than Carol's" since you cannot *compare* something which is entirely unlike anything else. See also entry for **Favorite.**

Where at. In a sentence like "I know *where* he is *at*," *at* is unnecessary and should be dropped: "I know *where* he is."

Which, who. Use *who* (or *that,* but never use *which*) to refer to persons. "Here is the man *who* [not *which*] is responsible."

While. *While* is mainly a conjunction of time: "I ran *while* I still had time." Do not overwork it to mean *and, but,* or *whereas*: "I loved roses, *but* [not *while*] she preferred daisies."

Whose, who's. A spelling mixup. See **Case, 7.**

Would of. This is a misspelling of *would've.* (See *should of.*)

Your, you're. *Your* is the possessive pronoun: "I am *your* friend." *You're* is the contraction of *you are*: "*You're* my best friend."

doc ——— DOCUMENTATION

To document is to identify your sources of information.

1. **[doc]** Document all outside sources of information used in your formal, research-based writing.
2. **[doc/cit]** Use the appropriate *style* (MLA or APA) for citations.
3. **[doc/bib]** Include a properly formatted bibliography.
4. **[doc/app]** See **Appendix on Documentation Styles** (p. 140) for examples of how to document the most common sources of information using either the MLA or APA style.

1. OVERVIEW: WHY DOCUMENT, WHAT TO DOCUMENT, HOW TO DOCUMENT doc

Why Document

The three primary goals of documentation are:

- Giving credit: to acknowledge the work of others is an ethical responsibility.
- Providing resources for your reader: to enable the reader to find your original sources is essential to the evaluation of your work and, in addition, gives your reader suggestions for further reading.
- Supporting your argument: to persuade your reader that your line of argument is valid, properly documented evidence may well be crucial.

What to Document

Document any information, opinion, or graphic that is not your own, but do not document items of general knowledge. For example, the statement, "Robert Kennedy was assassinated in 1968," does not need documentation because it is a well-known historical event. The following should always be documented:

- Direct quotations
- Paraphrases (presenting another's ideas in your own words)
- Any facts, data, figures, tables, or graphics that are copied or derived from another person's work
- An opinion or line of argument that you adopted from someone else's work
- Any work that helped you substantially in research for your paper

NOTE: If you are unsure, then it is best to document.

NOTE ON PLAGIARISM: The failure to identify a source so that the material appears to be presented as your own is called ***plagiarism***, or intellectual property theft. Plagiarism in a published document may result in legal action. In schoolwork it may result in disciplinary action, a failing grade, or both.

How to Document

Document your sources in accordance with a specific *style manual.* A style manual provides a set of rules for the placement and formatting of your *citations,* or acknowledgments of outside sources. Two of the most widely used documentation styles are those developed by the Modern Language Association (MLA) and the American Psychological Association (APA). The MLA style is widely used in the humanities, and the APA style is widely used in the social sciences. Some other disciplines have their own styles, and it is always wise to ask in advance which style to use for your paper. In most college writing courses you will be required to use either the MLA or APA style; consequently, these two styles are covered in this section. For extended treatment of these two styles, see the **Appendix on Documentation Styles.**

NOTE: Use one style in a paper consistently. Do not mix styles within a single paper.

2. IN-TEXT CITATIONS OF YOUR SOURCES OF INFORMATION doc/cit

- Provide an in-text citation for each reference to an outside source.
- Use the appropriate style for your citation.

A *citation* is a brief reference that identifies the source for the preceding information. In the MLA and APA styles it is also called an *in-text citation.* After quoting, paraphrasing, or in any way using an outside source, end your sentence with an in-text citation. A citation is short. The reader uses it to look up the full details of the source in the *bibliography,* which appears at the end of the paper. (See 3 below.)

NOTE: Footnotes and endnotes are seldom used for the documentation of sources. However, some documentation styles (such as that described in the *Chicago Manual of Style*) use footnotes or endnotes instead of in-text citations. There are handbooks available to guide you in these other documentation styles.

MLA Style

In the MLA style provide the author's last name (first name is optional), and the page or pages of the source in the first sentence in which you use the source material. Omit further citations within the same paragraph if the source of the material remains clear. As long as you continue to use the same source, provide only the page number for subsequent citations in the same paragraph. In general, the author's last name should be part of your sentence, and the page number(s) should appear in parentheses at the end of your sentence. If the author's name does not appear in your sentence, then insert, in parentheses at the end of your sentence, both the author's last name and the page(s).

EXAMPLES:

```
Ivan Klima writes: "throughout history there
runs a struggle between the powerful and the
powerless" (101).

Later, he says that after the war he
"devoured" literature about suffering (Klima
146).
```

NOTE: In the MLA style there is no punctuation between the author's name and the page number.

Multiple authors: If there are two or three authors, always provide all last names in a citation. If there are four or more authors, shorten the citation to the first last name followed by *et al.*

EXAMPLES:

`(Mast, Cohen, and Braudy 330-313).` [three authors]

`(Foley et al. 213).` [four or more authors]

EXCEPTIONS: If you are citing a work in whole, page numbers are not required:

`Henry VIII was Shakespeare's last play.`

[Since the author appears in the sentence and the work is cited as a whole, no extra citation is needed.]

If a citation does not clearly indicate which source in your bibliography is being cited, you will need to add additional information:

- If you are using two works by two different authors with the same last name, use first-name initials in the citation for clarity.
- If the work does not have an author or an editor, use as much of the title as necessary to identify the work—including at least one main word.
- If you are using more than one work by the same author, include as much of the title as necessary to identify the source—including at least one main word:

`Earlier in his career Foucault thought . . .`

`(Archaeology 123).`

`Later, he revised his ideas to include . . .`

`(Foucault, History 98-101).`

[Multiple works by the same author]

APA Style

In the APA style provide the author's last name and the year of publication in the first sentence in which you use the source material. Omit further citations within the same paragraph if the source of the material remains clear. In general, the author's last name should be part of your sentence, and it should be followed by the year in parentheses. If the name or names do not appear in the sentence, then insert the last name(s) and the year at the end of your sentence. Note, in the second example below, that in APA style you place a comma between the name and the year.

EXAMPLES:

`Goodwin and Jamison (1990) provided the most comprehensive reference to date on manic-depressive illness.`

`Manic-depressive illness is now usually called bipolar disorder (Goodwin & Jamison, 1990).`

[Notice that the period moves beyond the end of the citation.]

Multiple authors: If a source has one or two authors, list both last names every time you refer to it. For a source with three, four, or five authors, provide all last names in the first citation, but in later citations only the first author's last name followed by "et al." If the source has six or more authors, always use only the name of the first listed author followed by "et al." (If using "et al." does not clearly distinguish between sources in your bibliography, then give as many last names as needed to make the citation clear.)

EXAMPLES of shortened references:

`Foley, vanDam, Feiner, and Hughes (1990) discuss . . .` [First citation]

```
Later, the relative speeds of the computer
programs are compared (Foley et al., 1990).
```
[Shortened citation form for same source]

EXCEPTIONS:

- If you are citing a particular section of a text, include the page number(s).
- If you are using a source that is republished, include the date of original publication.

```
He later writes: "If left alone in the room,
each of you would probably involuntarily re-
arrange himself" (James, 1902/1982, p. 261).
```
[Always use quotation marks for direct quotations.]

3. CONSTRUCTING A PROPERLY FORMATTED BIBLIOGRAPHY doc/bib

1. At the end of your paper, list all your sources in a complete bibliography.
2. Use the proper format for your bibliography.

A *bibliography* is an alphabetized list of outside sources used in the research and preparation of your paper. Include a bibliography even if you use only one outside source. List all items you used including:

- Any item that is discussed or analyzed in your paper
- Any source that you quote, paraphrase (recast in your own words), or reference in your paper
- Any source for facts that are not a part of general knowledge

- Any source you used while researching your paper, even if not mentioned in your paper

Each documentation style has its own bibliography format. Follow its guidelines precisely. For styles other than MLA or APA, consult the appropriate handbook.

MLA STYLE: In the MLA style, the bibliography is called the "Works Cited" page.

APA STYLE: In the APA style, the bibliography is called the "Reference List."

OTHER STYLE: In other styles, the bibliography may be called the bibliography, reference page, or something similar. Usually, the index of a given style manual will give an appropriate reference within the book under "biliography" since this is a generic term.

See **Appendix on Documentation Styles** for detailed guidelines to help you prepare your bibliography for either style.

NOTE: Make sure that your in-text citations agree with your bibliography—and vice versa!

- Does each source you mention in an in-text citation have an entry in your bibliography?
- Do the details of the bibliography and in-text citations match?
- Finally, are all your references accurate? [Watch for slips such as writing ALBERT Einstein (the physicist) when you meant ALFRED Einstein (the music critic).]

4. DOCUMENTATION STYLES doc/MLA or doc/APA

See APPENDIX ON DOCUMENTATION STYLES (p. 140)

The Appendix is rich in examples of how to document the most common sources of information using either the Modern Language Association (MLA—Appendix 140) or American Psychological Association (APA—Appendix 147) styles.

-ed error — -ED ERROR IN -ED ENDINGS

Add *-ed* to indicate the past tense of the verb.

Often the past-tense *-ed* ending is not clearly heard in spoken English. Thus, it is easy to omit it in writing. Examples:

> **OMITTED:** *I hope* she would be there. [In many cases, of course, only the *d* is omitted rather than the full *-ed.*]
> **CORRECTED:** *I hoped* she would be there. [The *-ed* ending sounds like a *t*. A *t* may be hard to hear right after a *p*. The same is true for the *-ed* after the *k* in *asked* or after the *sh* in *wished.* No wonder you can easily omit it in writing.]
>
> **OMITTED:** She *try* to telephone him.
> **CORRECTED:** She *tried* to telephone him. [It is easy not to hear the *d* sound in *tried* right before the *t* in *to.* As to the spelling rule, note that verbs ending in a *consonant* plus *y—hurry* and *empty,* for example—change their *y* to an *i* and add either *-es* for the present tense or *-ed* for the past: *hurries, hurried, empties, emptied.* Notice, however, that

*There are three exceptions to this *ay*-verb rule: *Lay, pay,* and *say* regularly become *lays, pays,* and *says* in the present but change to *laid, paid,* and *said* in the past.

verbs that end in a *vowel* plus *y* undergo no such change: *play, plays, played.**]

OMITTED: *I use* to live in Mexico.
CORRECTED: *I used* to live in Mexico. [This is the most common error of this type. The *-ed* in *used* sounds much like a *t* and is not easy to hear before the *t* in *to*. The same is true for *suppose to*, which should be *supposed to*.]

NOTE: Do not drop the tense endings from past participles used as adjectives:

- The church boasted ornate, *stain*-glass windows. [Change "stain-glass" to "stained-glass".]

ELLIPSIS ell . . ./

In formal writing, the ellipsis—three double-spaced periods—is used only to show that you have omitted material from a *quoted* passage. (In creative writing—fiction, for example—the ellipsis is sparingly used to suggest emotion or to heighten suspense.) Do not use the ellipsis as a substitute for the dash or parenthesis.

Examples of the formal use of the ellipsis are as follows:

> Poetry turns all things to loveliness, it exalts the beauty of that which is most beautiful, and it adds beauty to that which is most deformed: . . . it subdues to union . . . all irreconcilable things.
>
> —Percy Bysshe Shelley

Use *four* double-spaced periods in the ellipsis when the material you have omitted ends a sentence:

> Branshaw Manor, says the author, "lies in a little hollow with lawns across it. . . . The immense wind, coming from across the forest, roared overhead." [The following was omitted from the *end* of the first quoted sentence, "and pine-woods on the fringe of the dip."]

To show that you are omitting one or more paragraphs from a quoted passage of prose or that you are cutting at least a full line from quoted poetry, use a full line of periods:

> What thou lovest well remains,
> . . . .
> What thou lov'st well is thy true heritage.
>
> —Ezra Pound

em ——— EMPHASIS

Give proper emphasis to the more important parts of the sentence and less emphasis to the less important parts.

Use the following methods for emphasizing important words or ideas.

1. ***Rearrange your sentence to give the important words and phrases their proper emphasis.*** The position of greatest emphasis is the *end* of your sentence. Next in emphasis is the beginning of your sentence.

   POOR EMPHASIS: We jammed into the car and started on our trip *in the morning, just after the sun rose.* [The italicized phrases are the least important elements of the sentence, but they are placed at the end, the position of most emphasis.]

PROPER EMPHASIS: In the morning, just after the sun rose, we jammed into the car and started on our trip. [The main clause, beginning *we jammed,* is now properly emphasized.]

2. ***Change the weak passive voice of the verb to the strong active voice.*** (See **Passive Voice.**)

UNEMPHATIC PASSIVE VOICE: At camp, many games *were played by the children* that were not played at home.
EMPHATIC ACTIVE VOICE: At camp, *the children played* many games that they did not play at home.

3. ***Underline a word or phrase for strong emphasis.*** Use sparingly. Underlined words in a manuscript appear in italics (*slanted type like this*) in print. Underline a word or phrase for strong emphasis only if you cannot achieve the emphasis by rephrasing or rearranging sentence parts. (See **Italics.**)

It is of course *possible* that all or any of our beliefs may be mistaken. . . . But we cannot have *reason* to reject a belief except on the ground of some other belief.

—Bertrand Russell

EXCLAMATION POINT— excl !/

Insert an exclamation point, or omit one if you have used one incorrectly. The exclamation point is used to express *strong* feeling. Do not overuse it.

PROPER USE: What a wonderful, wonderful day! [Exuberance.]
PROPER USE: Get out of here! [A brisk command.]

OVERUSE: The first baseman leaped up! He ripped the ball out of the air! The double play that followed was a cinch!

REVISION: The first baseman leaped up. He ripped the ball out of the air. The double-play that followed was a cinch. [These short, jabbing sentences are emphatic enough without exclamation points.]

frag — FRAGMENTARY SENTENCE

Change the sentence fragment to a complete sentence. You have written only a phrase or a subordinate clause, or some other *piece* of a sentence, but not a full sentence. If you can logically attach what you have written to the previous or the following sentence, do so. If not, expand your fragment into a full sentence by adding the missing element(s).

In certain types of creative writing, fragments are used effectively to suggest the frequently rapid and nongrammatical flow of thought, especially at emotional high points. But in formal, expository writing, where logic and calm are needed, sentence fragments are rarely appropriate.

The simplest sorts of fragments to correct are called *period faults.* They are sentences ended *too soon.* Usually a comma is needed in place of the faulty period. In the following examples, unjustifiable sentence fragments are in italics:

PERIOD FAULT: I do not have the steadiest hand in the world. *As you can see from my writing.*

The fragment is a subordinate clause that should be attached to the previous sentence by a comma. (For a further explanation, see **Comma, 3.**)

REVISION: I do not have the steadiest hand in the world, *as you can see from my writing.*

PERIOD FAULT: He spent some of his college years in Tucson. *A city whose weather is springlike eight months of the year.*

The fragment is an appositive. An *appositive* is a noun that renames or identifies a previous noun. In this case, the appositive *city* refers to the already named *Tucson.* Join the appositive to the first part of the sentence with a comma.

REVISION: He spent some of his college years in Tucson, *a city whose weather is springlike eight months of the year.*

In more complicated cases, fragments result when parts of a sentence are missing altogether:

FRAGMENT: She lectured on a number of occult subjects. *For example, about numerology.*

The fragment is a phrase. It is better style to make a complete sentence out of it than to add it to the previous sentence.

REVISION: She lectured on a number of occult subjects. For example, *she spoke* about numerology.

Better yet, however, would be to recast the entire idea: *She lectured on numerology and a number of other occult subjects.*

FRAGMENT: John would not make a good captain. *A good player, yes, but not always a good sport.*

The fragment here is a sentence part that needs a subject and verb. But the subject and verb were both omitted. Supply them:

REVISION: John would not make a good captain. *He is* a good player, yes, but not always a good sport.

hy -/ ——— HYPHEN

Insert a hyphen (-) where indicated. The hyphen is mainly used to connect words that are to be regarded as a unit of meaning: *fire-eater, helter-skelter, rabble-rouser.* (In many cases, usage is not generally agreed on even among dictionaries. In order to be consistent, choose one current dictionary and follow it.)

The following examples illustrate special uses of hyphens, such as preventing misreading and end-of-line word division.

1. Use a hyphen to connect modifying words before a noun when such words act as a unit of meaning:

- *an intelligent-looking face*
- *nineteenth-century history* [But not in *the history of the nineteenth century*]
- *a do-or-die attitude*
- *behind-the-scenes dealing* [But in the sentence "There were shady dealings going on behind the scenes," no hyphens are used because *behind the scenes* comes *after* the noun.]
- *four-, six-, and eight-cylinder cars* [Sometimes hyphens are *suspended* in a series before a noun.]

2. Use a hyphen to prevent misreading:

- *a foreign-car dealer* [Unless you mean *a foreign car-dealer*—a visiting Frenchman, for example, who sells cars in his native Paris.]
- *a small-appliance store* [Such a store could be very large, indeed, but nobody would believe it if you removed the hyphen!]
- They *re-covered* the chair. [Compare: They *recovered* the stolen chair.]
- I could think of one *more-expensive* gift to buy. [Compare: I could think of one more expensive gift to buy. (Here, "one more expensive gift" means one *additional* expensive gift.) The hyphenated example simply *compares* the cost of two gifts. The version without the hyphen means that *all* the gifts are in and of themselves expensive!]

3. Use hyphens for numbers between twenty and one hundred: *twenty-nine, sixty-two, eighty-eight.* (See **Numbers.**)

4. Use a hyphen for an end-of-line word division: *hyphenation,* not *hyphe-nation.* Divide words at the end of a full syllable. Follow the word divisions in a good dictionary, for example: sym•pa•thet•ic.

IDIOM — id

Use an acceptable idiomatic expression. An idiomatic expression, or idiom, is a linguistic form that occurs in one specific language and will not usually be found in any others. English is a language so rich in idioms that both native speakers of English and non-native speakers often misuse such expressions.

In English there are two areas that present problems involving idiomatic usage: (1) vocabulary, and (2) grammatical structure.

1. ***Vocabulary.*** Many expressions may seem illogical or nonsensical in the contexts in which they occur. Take the following sentence, for example: "He made money hand over fist." A native speaker is unlikely to have a problem with this remark, but a non-native speaker may not know that *to make money* means to *earn* money (not to manufacture it), and that to make money *hand over fist* means to make *lots* of money. Or take this sentence: "The fool wasted ten years before he came to his senses." The phrase *to come to (one's) senses*—finally to act more wisely—is immediately clear to most native speakers, but may not *make sense* (another idiom!) to a non-native speaker.

   Standard versus nonstandard idioms: Many idioms are perfectly acceptable in formal written usage as standard English, but there are many also that are *slang*, or *nonstandard*, and are labeled as such in any good dictionary. Use a desk-size college English dictionary (or a bilingual one, if preferable) to be sure of how an idiom is used, and write down idioms that you hear and find confusing. (For specific help with vocabulary, see also the brief sections in this book on **Diction, Jargon, Sexist Expressions, Slang,** and **Triteness,** and read the Glossary under **Diction.**)

2. ***Grammatical Structure.*** The idiomatic nature of English is not confined to matters of vocabulary but extends to grammatical structures as well. Some of the major problems related to idiomatic *grammatical* usage are found in this handbook in the sections **Article, Case, -Ed Error, Passive Voice, -S Error,** and **Tense.**

inc — INCOMPLETE CONSTRUCTION

Add the word or words necessary to complete the construction you now have.

PREPOSITION OMITTED: She was greatly interested and enthusiastic about the project.
REVISED: She was greatly interested *in* and enthusiastic about the project.

VERB OMITTED: The people were all interesting, and my vacation, in general, wonderful.
REVISED: The people were all interesting, and my vacation, in general, *was* wonderful. [The plural verb *were,* used with *people,* does not agree with the singular noun *vacation.*]

VERB OMITTED: We never have and never shall attack without provocation.
REVISED: We never have *attacked* and never shall attack without provocation. [The auxiliary verb *have* must be followed by *attacked.*]

RELATIVE PRONOUN OMITTED: He thinks that I am vicious and I can hardly wait to beat him to a pulp.
REVISED: He thinks that I am vicious and *that* I can hardly wait to beat him to a pulp. [What a difference a *that* makes! See a similar example under **Parallelism.**]

NOTE: See **Comparison** for other examples of incomplete constructions.

ITALICS ——— ital

Underline the word or passage indicated. When printed, underlined words appear in italic or slanted type, *like this.* If you underline (or italicize) words, you are asking the reader to pay greater-than-usual attention to them. Frequent underlining for the purpose of emphasis may backfire, however, for it

will no longer seem to signal what is unusual. Underline no more than is absolutely necessary—under normal circumstances, perhaps three or four times per page at most. There are four situations that call for underlining:

1. Underline the titles of books, magazines, newspapers, plays, and movies. Do not put quotation marks around them: *Neuromancer, Newsweek,* the *New York Times, Our Town, Star Trek VI.* (For further information see **Quotation Marks, 3.**)

2. Underline foreign words and expressions: *coup d'état, al fresco, chutzpah, persona non grata, touché!*

3. Underline words or letters if they are not used for their meaning, but as words or letters only.

   EXAMPLE: Add a *u* to *gaze* and you get *gauze.*

4. Underline words when strong emphasis is desired: "I'm leaving *now,*" she declared, "*not* tomorrow!" (See **Emphasis.**)

NOTE: Do not underline and do not put quotation marks around the title of your own essay or composition.

JARGON —— jarg

Do not use jargon, the highly technical language of professions and specialized interest groups, if you are writing for a general audience.

> **JARGON:** Digging in a long-buried *insula* in the old Roman camp, we uncovered a shard-filled *midden.*
> **REWRITTEN:** Digging in a long-buried *block of buildings* in the old Roman camp, we uncovered a shard-filled *heap of ancient refuse.* [In this version, the archeological terminology of the original is "translated" for the average reader.]

> **JARGON:** When I bit into the chocolate bar, I had an instant of *parageusia* and was sure that someone had spiked it with salt.
> **REWRITTEN:** When I bit into the chocolate bar, I had *a taste hallucination* for an instant and was sure that someone had spiked it with salt. [There's no need for the fancy psychological vocabulary.]

LOGIC —— log

Reread the marked passage and try to discover the flaw in your language or line of reasoning. You may have failed to argue your point convincingly. Your problem may be ineffective phrasing, ineffective thinking, or both. Think through your idea again, and if it still seems worth defending, try to present it more effectively.

Many of the problems in writing that damage your logic— the clear and convincing expression of your ideas—are dealt with in this handbook under the following headings: **Abstract Expressions; Ambiguity; Coherence; Incomplete Construction; Mixed Construction; Paragraph; Shift in Point of View; Transitions; Vagueness.**

If the problem lies mainly in the thought rather than in the expression, perhaps your difficulty falls into one of the following common categories of faulty logic:

1. Oversimplification
2. Overgeneralization
3. Appeal to authority
4. Appeal to emotion
5. Non sequitur

1. OVERSIMPLIFICATION log/simp

When you oversimplify, you make a statement that you want your readers to believe, but you give either inadequate evidence or too simple an explanation. Few statements can be absolutely proved, but you should present *good evidence* or *sound argument* to support your point of view. Good evidence consists of concrete examples and relevant facts and figures. A sound argument recognizes and tries to deal with the complexity of most issues:

EVIDENCE MISSING: The quality of student life on campus is poor. If conditions do not improve soon, many students may leave this university. [If no examples of the poor conditions that are claimed are given, the reader has no way of judging how true that first statement might be. Supply the evidence.]
IMPROVED (EVIDENCE SUPPLIED): The quality of student life on campus is poor. *In a recent student survey, 70 percent of those polled found the food at the student cafeteria "unacceptable," 62 percent found library personnel "unhelpful," and 84 percent found their dormitories "poorly managed."* If conditions do not improve soon, *the recently announced hike in tuition may drive* many students to leave this university.

COMPLEXITY UNRECOGNIZED: If families were more stable, no real narcotics problem would exist in this country. [This is an example of overly simple *cause-and-effect* reasoning. It is too simple to assume that unstable family situations *directly cause* narcotics addition. A revised statement would avoid such a narrow cause-and-effect claim.]
IMPROVED: The instability of the family is a *factor that contributes to* the narcotics problem in this country.

2. OVERGENERALIZATION log/gen

You are overgeneralizing when you allow *no exceptions* to your rule. Other than in the natural sciences, it is hard to find general statements that apply to absolutely every known case. Learn to *qualify* (to admit exceptions to) your statements. Be wary of using words like *always, never, all, none*:

OVERGENERALIZATION: College students are interested in partying first, studying last. [Learn to use *qualifying* words such as *some, many, sometimes, often.* As it stands now, *college students* implies *all* college students without exception.]
IMPROVED: *Many* college students are interested in partying first, studying last.

OVERGENERALIZATION: In the American family, the needs and desires of the children *always* come first. [If this were true, we would not hear of most child abuse.]
IMPROVED: In the American family, the needs and desires of the children *often* come first.

3. APPEAL TO AUTHORITY log/auth

In appeal to authority, you offer no stronger backing for your point of view than the word of a presumed authority on the subject. The word of established, well-recognized authorities may be good support for what we believe. But there is a lot wrong with relying on *false* authorities. The winner of the Indianapolis 500 is not necessarily the best authority on what beer to drink. Another problem is that even the best authorities sometimes contradict one another. What you should avoid, if at all possible, is relying on authority and nothing but authority for your beliefs:

APPEAL TO AUTHORITY: Jogging is one of the best exercises for ensuring good health and a long life. The governor's personal physician recently said so to reporters. [The social status of this politically connected physician has evidently impressed this writer tremendously. But surely there must be better evidence to be found for the benefits of jogging.]
IMPROVED: *Recently published evidence that joggers have a lower incidence of cardiac problems leads me to believe that* jogging is one of the best exercises for ensuring good health and a long life.

APPEAL TO AUTHORITY: Earth landings of flying saucers with alien beings in them have actually occurred, according to last week's *National Investigator,* and the government has for many years been suppressing news of them. [Suppose the *National Investigator* is a periodical

that is not known for objective, factual reporting. Although the statement *may* be true, the *authority* quoted in support of it will prevent most readers from giving it serious consideration.]
IMPROVED: *There are many unconfirmed reports that* Earth landings of flying saucers with alien beings in them have actually occurred, *and it is possible that* the government has for many years been suppressing news of them. [There is nothing wrong with making the most outrageous statements, so long as you properly *qualify* them—as the words *unconfirmed* and *possible* help to do.]

4. APPEAL TO EMOTION log/emot

In appeal to emotion, instead of using objective evidence and rational argument, you support your views by an appeal to your readers' emotions—to their prejudices, fears, or vanities:

APPEAL TO EMOTION: Why would anyone want Klinger as president of the university? He has twice been divorced, and his present wife is an alcoholic. [An appeal to many people's prejudices is substituted for an examination of Klinger's administrative credentials.]
IMPROVED: Why would anyone want Klinger as president of the university? *As we all remember, he was dismissed several years ago from his position as financial vice-president.*

APPEAL TO EMOTION: As almost any leading corporation executive will tell you, a Brooks Brothers suit is the only kind worth buying. [The appeal here is to the reader's desire for success and status and not to the quality of the suit itself.]
IMPROVED: *Although the price is high,* I prefer Brooks Brothers suits to *any others because of the excellent material and craftsmanship that go into them.*

5. NON SEQUITUR log/ns

The Latin term *non sequitur* means *it does not follow.* You move from one thought to another as if there were a logical connection between them, but there is none. (*Appeal to emotion* is a special case of *non sequitur.*) When one thought does not follow another, the reason may be, very simply, that you were hasty and left out a necessary bridge or transition. If you supply the transitional thought, the reader clearly sees the connection between the other two thoughts:

NON SEQUITUR: Because my next high school was much larger, we were allowed a longer lunch hour. [Something is missing here. What is the connection between the size of the school and the length of the lunch hour?]
IMPROVED: Because my next high school was much larger, *and the lines in the crowded cafeteria moved more slowly,* we were allowed a longer lunch hour.

(See **Coherence.**)

NON SEQUITUR: Because of an all-male student body, in class each student could be himself and act and speak naturally. [This is a very jumbled and misleading statement because at least two facts or ideas are left out.]
IMPROVED: Because the student body *at my high school* was all male, *showing off for the females was strictly an after-school distraction,* and in class each student could be himself and act and speak naturally.

MISPLACED MODIFIER — mm, mod

Put the misplaced word or phrase in a closer or clearer relation to the word it modifies.

A *modifier* is a word or group of words that adds descriptive detail to any of four types of words in a sentence: nouns, verbs, adjectives, and adverbs. In the sentence "He loves old cars," *old* is an adjective that modifies (describes) the noun *car.* In the sentence "They fought bravely," *bravely* is an adverb that modifies the verb *fought.* In the sentence "Be very careful," *very* is an adverb that modifies the adjective *careful.* In the sentence "She swims extremely well," *extremely* is an adverb that modifies the more important adverb *well.*

Sometimes a modifier is *misplaced.* It shows up too soon or too late in a sentence to connect unmistakably with the word it modifies. In such cases confusion may result, as in the following examples:

MISPLACED: They *only* wanted to steal what they needed. [Does the writer mean, "They would get what they needed *only through stealing*"? This is unlikely. The writer probably means, "They wanted to steal *no more than* what they needed." The proper placement of *only* can bring out exactly this meaning.]

CLEARER: They wanted to steal *only* what they needed.

MISPLACED: The emperor was just and kind to people *in his way.* [It sounds as if the emperor was kind to people who were obstacles in his path.]
CLEARER: *In his way,* the emperor was just and kind to his people.

MISPLACED: He fell while he was running *into a manhole.*
CLEARER: He fell *into a manhole* while he was running.

MISPLACED: The woman who was working *quickly* swallowed her lunch. [This is a case of a *squinting* modifier, one that can modify either of two words. Does *quickly* modify *working* or *swallowed*? Most probably it modifies *swallowed.* See also **Ambiguity.**]
CLEARER: The woman who was working swallowed her lunch *quickly.*

MISPLACED: We were awakened by the rattle of machine guns and sporadic bursts of rifle fire *that morning.*
BETTER: *That morning* we were awakened by the rattle of machine guns and sporadic bursts of rifle fire. [See similar examples under **Emphasis.**]

mx—MIXED CONSTRUCTION

Change one part of the sentence to make it match the rest. Your sentence begins with one construction or figure of speech, then shifts to another that cannot logically or structurally complete the sentence:

MIXED SENTENCE PARTS: By throwing the upper-right-hand lever is the way to stop the machine.

The two halves of this sentence do not fit together. The first part is an adverbial phrase that cannot serve as the subject. The second part, beginning with *is,* needs a noun as a subject. To correct the sentence, give it a subject in one of the following ways:

CORRECTION 1: *Throwing* the upper-right-hand lever *is* the way to stop the machine. [Leave out the *By*; then the first clause becomes a noun phrase and the subject of the verb *is.*]
CORRECTION 2: By throwing the upper-right-hand lever, *you can* stop the machine. [Change *is the way to* to *you can.* Now *you* is the subject of the sentence, and the adverbial clause *By throwing . . . ,* modifies the verb *can stop.*]

MIXED SENTENCE PARTS: *The reason* so few professors seek employment at this college *is because* we are located so far from any metropolitan center. [*Because* begins an adverbial clause that many writers wrongly put to work as a *noun.* Change the adverbial clause to a noun clause. *The reason . . . is because* becomes *The reason . . . is that.*]
CORRECTION: *The reason* so few professors seek employment at this college *is that* we are located so far from any metropolitan center. (See also *Reason is because* in "Words Often Misused: A Glossary" under **Diction.**)

MIXED SENTENCE PARTS: *Just because* she is rich is no reason to suppose she is happy.

This sentence has the same type of error as the previous one, except in reverse. This typical *just because* sentence forces an adverb clause—*because she is rich*—to serve as the subject. Subjects ought to be nouns, noun clauses, or other types of *noun* constructions. The simplest correction changes the adverbial clause into a noun clause: *Just because* becomes *The simple fact that* or *The mere fact that.*

CORRECTION: *The mere fact that* she is rich is no reason to suppose she is happy.

MIXED SENTENCE PARTS: Now, and not next month, is *when* we should write letters to our legislators.

The *is when* and *is because* constructions are similar: Adverbs are forced to do the job of nouns. Correct most *is when* sentences simply by striking out *is when* and switching the two halves of the sentence around:

CORRECTION: We should write letters to our legislators now, and not next month.

MIXED SENTENCE PARTS: In Shirley Jackson's short story "The Lottery," the danger of frozen ideas, as the narrator says, "no one liked to upset even as much tradition as was represented by the black box."

[When including a quotation within your own sentence, be sure that the quoted material flows smoothly into your own sentence structure. The test: remove the quotes and see if the whole sentence reads coherently, as if all the words were your own. In the example, the quotation has no grammatical connection with the rest. It is not immediately clear how the quotation is supposed to complete the thought. In the following revision, the connection becomes clear.]

REVISION: In Shirley Jackson's short story "The Lottery," the narrator expresses the danger of frozen ideas in saying, "no one liked to upset even as much tradition as was represented by the black box."

MIXED METAPHORS: The *wheels* of fate moved their grimy *hands.* [Do *wheels* have *hands*? As you can see, mixing metaphors (figures of speech) can result in an absurd image—funny, but not intentionally so.]
MIXED METAPHORS: A *tongue* of land jutted out from the *foot* of the cliff. [It is absurd to imagine a *foot* sticking out its *tongue.*]

num — NUMBERS

1. Spell out any figure that can be spoken in one or two words: ***thirty, fifty-five, three hundred, two million.*** **[NOTE: Hyphenate numbers between twenty and one hundred (*forty-two, seventy-one, ninety-six*).]**

2. Use numerals for any sum that must be expressed in three or more words: ***172, 307, 1002.***

EXCEPTION: Spell out figures that begin a sentence: *Three hundred sixty-eight* students work part-time.

¶ PARAGRAPH

Begin a new paragraph at the place marked. In most writing, the first line of a paragraph is *indented,* that is, begins several spaces to the right of where a line usually begins. In special cases, paragraphs can be indicated by blank lines between blocks of types or by introductory symbols such as an asterisk (*), bullet (•), or number.

The beginning of each new paragraph marks a new stage in the development of your essay. Just as sentences are the largest subunits of a paragraph, so paragraphs are the largest subunits of your essay. In formal writing, a paragraph does not begin and end just anywhere, as if the writer were playing Pin the Tail on the Donkey. The main idea, or topic, of an essay needs to be developed in each of its important *aspects,* and a new paragraph signals the shift to a new aspect of the discussion. (If your topic is broad enough and your essay long enough, each of the main aspects may be broken down into two or more sections, or paragraphs.) If, for example, you were writing a short essay on the types of teachers you have known and you ended up writing about five specific types, you might decide to write an introductory paragraph briefly mentioning your purpose and the five types you will treat. Then, you would probably devote *one paragraph* to a description (with examples) of *each type.* Your essay would, therefore, be divided into at least six paragraphs.

The main idea of each paragraph is usually found in one sentence called the topic sentence. (The theme of the whole essay is usually found in the first paragraph and is called the *thesis statement.*) A paragraph consists of a topic sentence, or main idea, plus a number of sentences that *develop* that main idea in some satisfactory way through a *logical argument,* or a series of *details* or *examples,* or any combination of these methods. The topic sentence is usually the first one in a paragraph, although it sometimes appears at the end, as a summary of details or examples that come before it.

There are three main organizational features to a well-structured paragraph:

1. *Unity,* the relevance of all sentences to the topic sentence. The avoidance of digressions—wandering from the paragraph topic.
2. *Development,* elaboration of the main idea with enough details, examples, arguments, and so forth to give the impression of full or adequate treatment.
3. *Coherence,* the connection between one sentence and the next in a logical pattern.

The use of linking devices is an essential means of achieving coherence. There are four main types of linking devices:

1. Transitional words or phrases
2. Repetition of key words or ideas
3. Pronouns
4. Demonstrative adjectives (*this, that, these, those*)

(For more information see **Coherence** and **Transitions.**)

> **EXAMPLE: One of the most significant changes in American family lifestyles in recent years has been the increase in the number of working women.** [Topic sentence] Each year more and more are joining the ranks of the employed in search of personal fulfillment and financial independence. *In fact,* [Trans. Phr.] over half of the adult female population is now in *the labor force.* [Rep. idea] *Of particular importance to the family lifestyle* [Rep. key words] is the fact that half of all children under eighteen now have working mothers. The Census Bureau reported that 45 percent of all *mothers* [Rep. key word] of preschool children are presently working. *That* [Demon. adj.] *figure* [Rep. idea]

is four times higher than it was just thirty years ago.

This increase in the number of working women has caused a redefinition of family roles. [Trans. sentence linking ¶s] **One important change is that husbands and children are expected to do more around the house.** [Topic sentence] In many households *children are expected not only to do* [Rep. key idea] the dishes and clean their rooms, but also to do the family grocery shopping, cook some of the meals, and help care for the younger children. *And because a working woman contributes to the family's economic welfare,* [Rep. key idea] husbands are increasingly sharing in what was once considered "woman's work": babysitting, cooking, and doing the laundry.

PARALLELISM ——— paral //

When two or more sentence elements are equally important in content and function, make them alike in grammatical form. *Parallelism* is the matching of sentence elements with one another: subject with subject, verb with verb, object with object, indirect object with indirect object, modifier with modifier. Usually, when these matching sentence elements appear in a pair or a series, they are connected with coordinating conjunctions:

> **NOT PARALLEL:** I enjoy going to movies, listening to music, and cards.
> **REVISION 1:** I enjoy *going* to movies, *listening* to music, and *playing* cards.

The verb is *enjoy,* and its objects are *going, listening,* and *playing.* Because these sentence elements are all direct objects, they should be in the same form. In this case the objects should all be gerunds: *-ing* words used as nouns. Note that each gerund in the example is followed by a word or words (going *to movies*) that, together with the gerund, make a *gerund phrase.*

> **REVISION 2:** I enjoy *movies, music,* and *cards.*

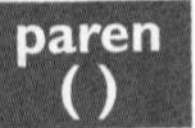

This series is also parallel because the three objects—movies, music, and cards—are all nouns. The gerund constructions have been removed.

> **NOT PARALLEL:** The opinion of one off-beat columnist is that the mayor is adept at underhanded dealings profitable only to himself, and we should therefore throw him out of office.
> **REVISED:** The opinion of one off-beat columnist is *that* the mayor is adept at underhanded dealings profitable only to himself and *that* we should therefore throw him out of office.

The second *that* makes it clear that the opinion that follows is solely the columnist's and does not belong to the writer of the sentence. You see how important clear parallelism can be.

NOTE: The comma after "himself" is now omitted because "and" does not introduce a new main clause. (See **Comma, 1.**)

> **NOT PARALLEL:** She always plays the piano with ease, with confidence, and takes pleasure in it.
> **REVISED:** She always plays the piano *with ease, with confidence,* and *with pleasure.*

In the revised sentence we have a series of parallel prepositional phrases. Because the three phrases modify the same verb—*plays*—they should be parallel.

paren ( )———PARENTHESES

Use parentheses to enclose material that is clearly additional commentary or detail and not essential to the meaning of the sentence. Whatever is enclosed in parentheses appears relatively unimportant to the reader. Use parentheses sparingly; never use them when you can use commas instead.

Effective Use of Parentheses

- I walked right up to him (no one was with him at the time) and told him what we had decided.

- Last week she came up with a brilliant new idea (the seeds of it had been germinating in her mind for months) only to see it rejected as absurd by the committee.

Ineffective Use of Parentheses

- His brother told him (John) not to annoy him (Allen) any more.

You should not compensate for poor reference of pronouns by using explanatory parentheses. The sentence needs rewriting to achieve a smooth style and clear meaning: *His brother told John not to annoy Allen anymore.*

- Some critics think that atmospheric pollution (not the population explosion) is the more serious challenge to human survival.

Commas are better than parentheses to set off this relatively important part of the sentence.

NOTE: Do not place a comma before a parenthesis: I had eaten too much, (two breakfasts that day!) and felt guilty about flouting my diet. [Omit comma after *much.*]

PASSIVE VOICE ——— pass

Make your writing more direct by changing the verb from the passive voice to the active voice. (See *Emphasis*.)

In a sentence in the *active voice*, the subject is *doing* something:

> **ACTIVE:** The referee blew the whistle. [The subject is *referee*.]

In a sentence in the *passive voice*, something is being *done to* the subject:

> **PASSIVE:** The whistle was blown. [The subject is *whistle*.]

Notice how the following sentences are more direct in the active voice:

> **WEAK PASSIVE VOICE:** With the changing of seasons there comes a change in the type of clothing *to be worn.*
> **DIRECT ACTIVE VOICE:** With the changing of seasons there comes a change in the type of clothing *people wear.* [Even better: *With the changing of the seasons people change the type of clothing they wear.*]

> **PASSIVE:** In the fall, cotton clothes *are stored* away by families.
> **ACTIVE:** In the fall, *families store away* their cotton clothes.

> **PASSIVE:** This book *should be read by all of you* as soon as it *can be bought* in paperback.
> **ACTIVE:** *You should all read* this book as soon as *you can buy* it in paperback.

Effective Uses of the Passive Voice

You may use the passive voice if the receiver of the action, or the action itself, is more important than the doer:

- The mayor of New York City *was bitten* by a horse today. [The *receiver* of the biting, the mayor of New York, gets top billing in this sentence.]

- The restaurant fire was started by a dripping panful of hot grease. [The *action,* the fire, is of more interest than who or what started it.]

You may use the passive voice if the doer of the action is unknown:

- Some years ago a treasure-laden Spanish galleon *was recovered* not far from the coast of Florida. [The writer may not know who recovered it or simply may not find that detail worth mentioning.]

Except in cases when the passive voice is needed, a careful writer will avoid passive voice constructions because they often lead to weak, roundabout, wordy sentences.

PRONOUN REFERENCE—pro

Make the pronoun you have used refer clearly to a previous noun (its antecedent).

1. DANGLING PRONOUNS

Certain pronouns, especially *which, this,* and *it,* are sometimes found to be *dangling* in a sentence because they simply have no antecedent—no noun—within reach for them to refer to. Dangling pronouns usually mean well. Often, they try to refer to a whole previous idea, but the idea has not been cast in the form of a noun. To revise the idea, restate it in the form of a noun:

UNCLEAR REFERENCE: He felt extremely angry toward her, which made it difficult for him to speak to her.

The dangling pronoun *which* is used—impossibly—to refer to the whole preceding main clause. It lacks an *antecedent,* a *noun* to refer to. Revise the idea in any way that eliminates the dangling *which.*

REVISION 1: He felt extremely angry toward her. His attitude made it difficult for him to speak to her. [This solution turns one sentence into two.]
REVISION 2: His extreme anger toward her made it difficult for him to speak to her. [An economical one-sentence solution!]

UNCLEAR REFERENCE: Anthony played varsity football, chaired the history club, spoke fluent Italian, and drew cartoons for the college newspaper. *This* explains why the student body elected him president.

The pronoun *this* is forced to refer to the whole preceding sentence. To revise, combine the pronoun *this* with a noun, or perhaps a more descriptive noun phrase, that sums up the ideas of the whole previous sentence.

REVISION: Anthony played varsity football, chaired the history club, spoke fluent Italian, and drew cartoons for the college newspaper. *This record of all-round excellence* explains why the student body elected him president.

UNCLEAR REFERENCE: How can you not be happy to see *the leaves appear in the spring* and the difference *it* makes in everything around you? [The pronoun *it* awkwardly refers to the whole previous clause, *the leaves appear in the spring.*]
REVISION: How can you not be happy to see the leaves in the spring and the difference their appearance makes in everything around you? [Sometimes, through rewriting, you can elimate a pronoun-reference problem simply by getting rid of the troublesome pronoun (*it*, in this case)!]

UNCLEAR REFERENCE: *It said* in last week's newspaper that a major reduction in Pentagon spending could initiate a recession.

Similar to this kind of sentence is the one that begins *They said on the radio that. . . .* The *it*—or *they*—implies a subject that is not stated. In formal writing, avoid this kind of construction. A simple revision uses *I read* or *I heard* to begin such a sentence.

REVISION: *I read* in last week's newspaper that a major reduction in Pentagon spending could initiate a recession.
BETTER: Last week's newspaper voiced the opinion that a major reduction in Pentagon spending could initiate a recession.

2. AMBIGUOUS PRONOUNS

Some pronouns refer *ambiguously* to either of two previous nouns. Pronouns should refer to one noun and one noun only. Your job is to remove the ambiguity. Reduce your pronoun's workload to one noun that it clearly refers to. (See **Ambiguity.**)

AMBIGUOUS REFERENCE: When John spoke to Peter, he said he doubted that *he* would be invited to the party. [Does the last *he* refer to John or Peter?]
REVISION: When John spoke to Peter, he said, "I doubt that I will be invited to the party." [The use of direct quotation is the simplest way to solve this sort of reference problem.]

AMBIGUOUS REFERENCE: Even if the salesperson persuades you to purchase an item, do not sign anything. Ask her to leave the contract with you so that you and your wife may read it thoroughly. If *she* will not *do it,* show her the door. [In the last sentence, the ambiguous use of *she* and *it* can lead to a rather amusing interpretation.]
REVISION: Even if the salesperson persuades you to purchase an item, do not sign anything. Ask her to leave the contract with you so that you and your wife may read it thoroughly. If the *salesperson* will not *leave it with you,* show her the door. [It is not enough to change *she* to *salesperson.* The dangling *it* in *do it* has no noun to attach to. It is made to refer to either of two verb-object combinations: *leave the contract* or *read it.* To clarify, change *do it* to *leave it with you.*]

qm ?/———QUESTION MARK

Do not omit the question mark or use it unnecessarily.

NOTE: Your instructor may sometimes use this symbol [?] not to point out a punctuation error but to question the logic or clarity or factual precision of a passage.

Use a question mark only after a *direct* question.

> **ERROR:** “Will you take out the garbage,” he requested. [Replace the comma after *garbage* with a question mark.]
> **ERROR:** Would I ever succeed, I asked myself. [Replace the comma after *succeed* with a question mark.]

Do not use a question mark after an *indirect* (or reported) question.

> **ERROR:** He requested that I take out the garbage? [Replace the question mark with a period. This is not a direct question but a report *about* a question.]
> **ERROR:** I asked myself if I would ever succeed? [Replace the question mark with a period. Note that if a *direct* question followed *I asked myself,* it would be *Would I ever succeed?*]

quot “/”—QUOTATION MARKS

Use quotation marks to set off (1) directly quoted words, (2) words used in an unusual way, and (3) the titles of subunits of a book or magazine—a chapter, story, poem, and so on.

The following instructions show when and how to use a pair of quotation marks (“ ”).

WHEN TO USE QUOTATION MARKS

1. ***Use quotation marks to enclose a passage of directly quoted words:***

- The flight attendant said, “Fasten your seat belts, please.”
- In his *Essay on Man,* Alexander Pope writes, “Hope springs eternal in the human breast.”

CAUTION 1: Do not use quotation marks for *indirect* quotations. An indirect quotation is a second-hand report of what someone said, and it is often introduced by the word *that:*

MISUSE: My brother said that “he was unhappy about the outcome.”
REVISION: My brother said that he was unhappy about the outcome. [Remove the quotation marks. Quotation marks, however, could be used around the words in this sentence that are actually quoted: My brother said that he was “unhappy about the outcome.” Or a complete direct quotation could be used: My brother said, “*I am* unhappy about the outcome.”]

CAUTION 2: In quoted dialogue, start a new paragraph with every change in speaker:
CLUTTERED: “Do you realize,” she said, “that we'll next be seeing each other on Friday the thirteenth?” “Do you think I'm superstitious, Lisa?” he said. “Of course not, Jim, but what made you shudder like that?”

REVISED: “Do you realize,” she said, “that we'll next be seeing each other on Friday the thirteenth?”
“Do you think I'm superstitious, Lisa?”
“Of course not, Jim, but what made you shudder like that?”

Keep identification of the speaker, such as *he said* or *she said,* outside of the quotation marks:

- “I suppose,” *she remarked,* “that success comes only with time.” [Because the quoted passage is one complete sentence, the interrupting words are set off by commas and not followed by a period or semicolon.]

- “I understand the plan,” *Jim said.* “I think it might work.” [In this case, two separate sentences are quoted. *Jim said* must be followed by a period, for it marks the end of one quoted sentence.l

When quoting long prose passages (five lines or more), do not use quotation marks. Instead, indent the entire passage five spaces to the right (a few more spaces for the first line of a paragraph) and single space if you are typing. The following example is from a student research paper.

> The author sums up in a nutshell the basic conditions that have shaped Spain’s cultural development:
>
> > Spain is a world apart from the rest of Europe, separated by climatic differences and isolated in time as well as in space. Bounded by water on three sides, and on the fourth cut off by the barrier of the Pyrenees, she was for three hundred years, from the VIIIth to the XIth centuries, virtually under the domination of an oriental power—that of the Moors, whose culture was not only more advanced than that of any part of Europe, but also profoundly different from any European civilization.
> >
> > —Enriqueta Harris, *Spanish Painting*

When quoting one or two lines of poetry, follow the example for a directly quoted passage:

- Wallace Stevens is playing a musical joke when he writes, “Chieftain Iffucan of Azcan in caftan/ Of tan with henna hackles, halt!” [Note the slash used to show the line end.]

When quoting more than two lines of poetry, do not use quotation marks. Simply indent the whole passage as for long prose quotations, single space if typing, and reproduce the original as it stands.

2. ***Use quotation marks to emphasize words used in an unusual sense:***

- In the printing trade, an engraved plate is called a "cut."

NOTE: Do not use quotation marks to ask acceptance for lazy, imprecise language of your own.

> **MISUSE:** In spite of Jerry's "goofing off," he was generally regarded as a very "sharp" character.

3. ***Use quotation marks to set off the titles of chapters of a book; episodes of a TV or radio series; and the titles of articles, short stories, and poems published as part of a complete book, magazine, or newspaper:***

- One of my favorite short stories is Gabriel Garcia Marquez's "A Very Old Man with Enormous Wings."

The title of the whole book is underlined, or italicized, whereas the chapter title is in quotation marks:

- When you read *Sister Carrie,* pay careful attention to the first chapter, "The Magnet Attracting: A Waif Amid Forces."

The *series* title of a TV or radio show is underlined, whereas the title of an individual episode is in quotation marks:

- Last week I watched an excellent program called "The Winged World" on *Best of the National Geographic Specials.*

NOTE: Do *not* use quotation marks around the title of your *own* essay. Do not underline your own title, either. To set it off clearly from the beginning of your essay, just skip a line or two between your title and your first paragraph.

How to Use Quotation Marks with Other Punctuation

1. ***Place periods and commas*** inside ***closing quotation marks:***

- Shakespeare said, “Unquiet meals make ill digestions.”
- Francis Bacon remarked that “the monuments of wit survive the monuments of power,” and I wholly agree with him.

2. ***Place colons and semicolons*** outside ***closing quotation marks:***

- We had arrived at “the moment of truth”: the matador’s extending his sword for the finishing stroke.
- I know that “to err is human”; yet fifteen errors in one ball game is too much to forgive.

3. ***Place question marks and exclamation points*** inside ***closing quotation marks*** only ***if they are a part of the quoted passage:***

- She asked him, “Is dinner ready?”
- He shouted, “Advance or I’ll fire!”

Place such marks outside the quotation marks if they are *not* part of the quoted passage:

- Did I just hear you say, “Dinner is ready”?
- Stop saying “yes”!

In direct quotations, avoid the unnecessary comma or period with a closing quotation mark:

> **AVOID:** “Well, well!”, he said. [Remove the comma.]
> **AVOID:** “Well, well!,” he said. [Remove the comma.]

BETTER: "Well, well!" he said.

AVOID: He asked, "What's your name?". [Remove the period.]
BETTER: He asked, "What's your name?"

4. ***Use single quotation marks (' ') to set off a quotation within a quotation:***

- "When Caesar said 'I came, I saw, I conquered,'" my history teacher declared, "little did he know that he had invented the telegram."

If you quote an essay title that quotes another title, use single quotation marks for the quoted title within the title:

- I just read a journal article about Robert Frost and his poem "Two Tramps in Mud Time." The title of the article is "Frost and the Work Ethic: 'Two Tramps in Mud Time.'"

In the journal, the title has no quotation marks around it, but the title of the poem it quotes is set in *double* quotation marks:

- Frost and the Work Ethic: "Two Tramps in Mud Time."

REPETITION — rep

Do not awkwardly repeat the same word or idea you have used before. (See *Wordiness.*) The more serious problem is the repetition of the same idea—whether in similar form or not—throughout the whole length of your essay. (See *Paragraph.*) Such repetition suggests that you have little to say but feel pressured to fill up space. The following examples show awkward uses of repetition and correction of the awkwardness:

REPETITION: A cool breeze was *blowing,* and the brownish gold leaves were being *blown* about by the wind.

Change *blown* to *swept,* or find another good synonym. Sometimes, as in this case, the sentence would be better if it were condensed. There is no need to refer to the *breeze* again, even by the synonym *wind:*

REVISION: The brownish gold leaves were being swept about by the cool breeze.

REPETITION: The air was too cold, *and* while I was asleep it chilled me, *and* when I awoke my bones felt stiff. [Eliminate one of the *and*'s.]
REVISION: The air was too cold. While I was asleep it chilled me, and when I awoke my bones felt stiff.

REPETITION: I want to earn a college degree *because* I would like many options to remain open to me in the future *because* rapid technological changes leave people of meager education at a great disadvantage. [Avoid the *because . . . because* construction.]
REVISION: I want to earn a college degree because I would like many options to remain open to me in the future. Rapid technological changes leave people of meager education at a great disadvantage.

REPETITION: The steam could be seen rising from the radiator. *The steam* turned to frost on the windowpane. [Delete the second *steam* and combine the two sentences.]
REVISION: The steam rising from the radiator turned to frost on the windowpane.

REPETITION: He kept changing his mind *and failed to stick with a single idea he came up with.*
REVISION: He kept changing his mind. [Avoid needlessly repeating an idea.]

Repetition is not always bad. (See **Parallelism.**) Repetition of words or ideas can be used deliberately to arouse emotion or aid memory. We find such conscious repetition in the most varied kinds of communication, from commercial advertising and political propaganda to poetry and oratory. The last major speech of Martin Luther King Jr. is unforgettable because of the mounting repetitions, "I have a dream . . ." And in the Bible, Ecclesiastes 3:1–3, the repetitions sound with a poetic beat: "To every thing there is a season, and a time to every purpose under the heaven. A time to be born, and a time to die; a time to plant, and a time to pluck up that which is planted; a time to kill, and a time to heal . . ."

RUN-ON SENTENCE——ro

Do not run one sentence into the next. Separate them with proper punctuation or a coordinating conjunction. End the first sentence with a period or other end-of-sentence punctuation such as a question mark or exclamation point and begin the next sentence with a capital letter. A semicolon will sometimes work better than a period or conjunction. (See *Semicolon, 2.*) In any case, do *not* join two sentences with only a comma. Although this comma error is sometimes called a run-on sentence, it is usually referred to as a *comma splice* (see *Comma Splice*):

> **RUN-ON:** Virtual Reality is a life-like three-dimensional computer world it inhabits what is called "cyberspace."
> **REVISION:** Virtual Reality is a life-like three-dimensional computer world. It inhabits what is called "cyberspace."

> **RUN-ON:** She put on her bathing cap then she plunged into the water.
> **REVISION 1:** She put on her bathing cap. Then she plunged into the water.

Although a period after *cap* is correct, a semicolon would probably be better. (Look up the use of semicolons before

conjunctive adverbs such as *then, however,* and *therefore* under **Comma Splice.**)

> **REVISION 2:** She put on her bathing cap, then plunged into the water.

By eliminating the second *she* and adding a comma after *cap,* we are left with a single, smooth sentence.

> **RUN-ON:** The snow fell all night in the morning the air was crystal clear.
> **REVISION 1:** The snow fell all night. In the morning the air was crystal clear.
> **REVISION 2:** The snow fell all night, *but* in the morning the air was crystal clear.

Sometimes a comma and a coordinating conjunction such as *but* or *and* make a smoother sentence. (See **Comma, 1.**)

NOTE: When sentences are short, they may be joined with a coordinating conjunction and no punctuation:

- He ran *and* she hid.

SEMICOLON — semi ;/

1. Use the semicolon to connect two main clauses when they are closely related in idea.
2. Use the semicolon to separate sentence elements equal in rank when they contain commas.

1. CONNECTING MAIN (INDEPENDENT) CLAUSES

Use the semicolon to connect two related complete sentences, generally when they are not connected with a coordinating conjunction such as *and, but, for, or, nor:*

- It is not so much the threatening weather that concerns me; it is the dilapidated condition of the ship. [The ideas are closely related.]

Also use the semicolon between main clauses connected by certain conjunctive adverbs, such as *however, therefore, then, similarly, likewise.* (See **Comma Splice.**)

- The written part of the exam does not faze Cindy; *however,* to become a state trooper she must also be able to benchpress half her weight.

Do not use the semicolon between a main clause and a phrase or subordinate clause:

> **AVOID:** I do not like to eat orange peels; although I admit that in marmalade they are quite good. [Remove the semicolon.]
> **BETTER:** I do not like to eat orange peels although I admit that in marmalade they are quite good. [No punctuation is normally needed between a main and a subordinate clause. See **Subordination** for a list of *subordinating conjunctions* that will help you to identify subordinate clauses.]

2. SEPARATING EQUAL ELEMENTS

Use the semicolon to show the main divisions in equal sentence elements containing commas:

- I introduced him to Alicia Soto, the president; Leroy Wilson, the vice-president; and Herb Dunn, the treasurer.

-s error — -S ERROR IN -S ENDINGS

Add an *-s* ending to your verb or omit an *-s* ending from your verb.

Standard English contradicts the speech habits of many people when it comes to the endings of verbs in the present tense:

NONSTANDARD: He *like* to play baseball. [The -*s* is missing from *likes*.]
STANDARD: He *likes* to play baseball.

NONSTANDARD: My cousin *work* all night and *sleep* all day.
STANDARD: My cousin *works* all night and *sleeps* all day.

NONSTANDARD: I *limits* myself to two hours of TV a day. [The -*s* should be omitted from *limits*.]
STANDARD: I *limit* myself to two hours of TV a day.

NONSTANDARD: They just never *stops* talking.
STANDARD: They just never *stop* talking.

NONSTANDARD: Today's women *appreciates* the importance of a good college education.
STANDARD: Today's women *appreciate* the importance of a good college education.

When do you keep the -*s* and when do you omit it? The pattern is simple:

- Present-tense verbs following the pronouns *he, she,* and *it* end in -*s*. (He goes; she drives; it smells.)
- Present-tense verbs following the pronouns *I, you, we,* and *they* do not end in -*s*. (I go; you drive; we try; they smell.)
- In the same way, present-tense verbs following nouns that substitute for *he, she,* and *it* end in -*s*: John (instead of *he*) goes; Ella (for *she*) drives; the barn (for *it*) smells. Also, present-tense verbs following nouns that substitute for *we* and *they* do not end in -*s*: Robert and I (instead of *we*) work hard; apples (for *they*) taste good.

NOTE: For a more detailed grammatical explanation of this pattern, see **Agreement, 1.**

sexist——SEXIST EXPRESSION

Avoid terms that refer to only one sex if they *unfairly* exclude the other. Whenever possible, choose pronouns and occupational titles that reflect equal treatment of the sexes.

1. SEXIST PRONOUNS

A special problem in the use of pronouns may occur when the following words are used as antecedents: *each, every, everyone, everybody, everything, someone, somebody, anyone, anybody, no one, nobody, either, neither, another.* Although they may occur in a sentence as antecedents, these words are themselves singular pronouns and should be referred to by singular pronouns.

> **EXAMPLE:** *Each* of us knows *his* job. [The use of *his* assumes that us consists entirely of males.]
> **EXAMPLE:** *Each* of the women presented *her* opinion.

NOTE: When any of these antecedents stands for a group of both men and women, you have several options:

A. You may use a double pronoun: Each of us knows *his or her* job well. Because double references can get cumbersome if overused, try the following alternate ways to represent both males and females in a group.

B. Leave out the pronouns entirely where possible:

> **CORRECT:** Everyone made *his* or *her* presentation.
> **BETTER:** Everyone made a presentation.

C. Use plurals where possible:

CORRECT: Everyone made *his or her* presentation.
BETTER: *All managers* made *their* presentations.

D. When omitting the pronouns or using the plural is not possible, write or rewrite to avoid implying that the group you are referring to, which actually includes both men and women, is composed exclusively of men:

MISLEADING: Anyone, if *he* works hard enough, can succeed.
BETTER: Anyone *who* works hard enough can succeed.

E. In passages requiring repeated use of a pronoun referring to someone *in general,* not to a specific person of known sex, it has become accepted practice to use *feminine* pronouns throughout at least the *first* such passage. (You may continue to do so throughout your essay, or you may prefer to *alternate* passages that use feminine pronouns with those using exclusively masculine pronouns—provided you do not confuse the reader.)

EXAMPLE (passage not about a specific individual): The writer who wishes to succeed in her chosen field needs more than just talent. She needs enormous self-confidence and the ability to take a stream of rejection slips in stride. Beyond that, she needs perseverance—maintenance of a rigorous work schedule through the years she may have to wait for recognition.

NOTE: If you remember to use *plurals* wherever possible (see C above), you can avoid choosing a sex-specific pronoun altogether and write the above paragraph as follows:

Writers who wish to succeed in *their* chosen field need more than just talent. *They* need enormous self-confidence and the ability to take a stream of rejection slips in stride. Beyond that, *they* need perseverance—maintenance of a rigorous work schedule through the years *they* may have to wait for recognition.

2. SEXIST OCCUPATIONAL TITLES

Avoid labeling an occupation as belonging particularly to one sex. If you use the word *foreman,* for instance, you may suggest to your readers—even without meaning to—that you assume anyone in charge of a work crew to be a male. Although such language is in common use, many people consider it sexist, implying prejudice against one of the sexes. Instead of *foreman* you might use a sexually neutral term like *supervisor* or *section head.* Some common sexist expressions and their nondiscriminatory alternatives follow:

Sexist	Alternative
Businessman	Business executive, business owner, business person, company head, manufacturer, wholesaler
Chairman	Chair, chairperson
Congressman	Representative, member of Congress
Fireman	Firefighter
Foreman	Supervisor
Garbageman	Sanitation worker
Housewife	Housekeeper
Mailman/postman	Postal clerk, letter carrier
Manpower	Workers, work force, personnel
Marksman	Sharpshooter
Policeman	Police officer
Repairman	Service technician
Salesman/saleswoman	Sales clerk, salesperson, sales representative
Weatherman	Meteorologist, weather forecaster
Workman	Worker, laborer, employee

SHIFT IN POINT OF VIEW —— shift

1. **Make the pronouns in your sentences agree in *number* and *person.***
2. **Make the verbs in your sentences agree in *tense, mood,* and *voice.***
3. **Do not carelessly present as your own statements those that should be attributed to others.**

1. PRONOUN SHIFTS

Avoid Shifts in Number

A shift in pronoun number occurs when you shift between singular and plural pronouns while referring to the same noun:

> **SHIFT:** During the sixties women gained more control over their roles in society than they had previously possessed. Now is no time for *her* to give up *her* struggle for equal rights.
> **REVISED:** During the sixties women gained more control over their roles in society than they previously possessed. Now is no time for *them* to give up *their* struggle for equal rights.

(See **Agreement, 2.**)

Avoid Shifts in Person

The personal pronouns are classified as *first-, second-,* and *third-person* pronouns (singular or plural). (See **Case.**) Examples of first-person pronouns are *I, me, we, our.* Second-person pronouns include *you, your, yours.* Among the more common third-person pronouns are *he, she, it, one, they.* Avoid needless shifts in person not only within the same sentence but also

from sentence to sentence and from paragraph to paragraph throughout the entire essay:

> **SHIFT:** If *one* stops to watch them work, *you* are greeted with a smile.
> **REVISION 1:** If *one* stops to watch them work, *one* is greeted with a smile. [Avoid frequent use of the impersonal *one.* It can make your style seem stiff.]
> **REVISION 2:** If *you* stop to watch them work, *you* are greeted with a smile.

2. VERB SHIFTS

Avoid Shifts in Tense

A verb's tense tells your reader when the action takes place. (See **Tense.**) If you shift unnecessarily, for instance, from the past to the present, you may confuse your reader. Avoid unnecessary shifts in tense not only within the same sentence but also from sentence to sentence and from paragraph to paragraph throughout an essay:

SHIFT: He *rushed* to catch his train but *misses* it by half a minute. [A needless shift from the past—*rushed*—to the present—*misses.*]
REVISED: He *rushed* to catch his train but *missed* it by half a minute.

Avoid Shifts in Mood

English has three verb moods: the *indicative,* the *imperative,* and the *subjunctive.* Most of the time, we speak and write in the indicative mood—the forms used for making declarative statements such as "I *work* on weekends." We use the *imperative* mood when we issue commands (to an implied you): *"Work faster."* The most common mood-shift *error* is from the imperative to the indicative:

SHIFT: Be sure to visit the science exhibition, and then you *should go* to the art show. [*Be* is the imperative; *should go* is the indicative.]
REVISED: *Be* sure to visit the science exhibition, and then go to the art show. [Both verbs, *be* and *go,* are now in the imperative mood.]

Occasionally, we use the *subjunctive* mood—when we discuss possibilities instead of facts and when we use certain stock phrases like "*Be* that as it may." Of the very few uses of the subjunctive in English, here is a typical one—proposing a condition contrary to fact—involving the use of the past tense of "to be":

INCORRECT: If Stephen King *was* actually a group of authors, I could understand his enormous productivity.
CORRECT: If Stephen King *were* actually a group of authors, I could understand his enormous productivity.

Avoid Shifts in Voice

Avoid changing from the active voice to the passive voice when your subject is still active. (See **Passive Voice.**) Often there is no need to change the subject from clause to clause within the same sentence:

> **SHIFT:** *We* predicted the results more easily after the *margin of error* had been reduced. [Who reduced the margin of error?]
> **REVISED:** *We* predicted the results more easily after *we* had reduced the margin of error.

3. SHIFTS IN THE ATTRIBUTION OF STATEMENTS

Be Clear About Who Is Saying What

Make sure that when you present the thoughts of others, the reader cannot take them as your own statements or opinions. Only a few words of attribution are needed for clarification:

> **SHIFT:** In Eudora Welty's "Petrified Man," women are portrayed as victims of Hollywood-inspired fantasy. *The real men in their lives are all hindrances, millstones, and rapists.* [The italicized sentence appears to be the opinion of the student who wrote this passage.]
> **REVISED:** In Eudora Welty's "Petrified Man," women are portrayed as victims of Hollywood-inspired fantasy. *The female characters feel that* the real men in their lives are all hindrances, millstones, and rapists. [Addition of the italicized words makes it clear that the opinion that follows is not the student's own but belongs to the characters in Welty's story.]

SLANG —— slang

Avoid the use of slang—catchy, colorful words or phrases currently in use by seemingly everyone, especially the media—when writing formal English.

Slang expressions tend to be popular for a few years but soon sound dated and are replaced by other, equally glittery coinages. Many years ago people used to say "copasetic"—a word recognized only by specialists today—to express the same kind of mindless approval as is now generally rendered by "awesome." Some slang, however, fails to die out and enriches the language permanently as a *colloquialism.* Respectable as even colloquialisms are in *informal* writing, though, they still sound largely out of place in *formal* written English. The problem for any writer is how to sound fresh and interesting without using slang or any other kind of overused language. (See hints for improving your style under **Triteness** and **Diction.**)

> **MISUSE OF SLANG IN A FORMAL CONTEXT:** The president decided to *blow off* his appointment with the British prime minister when the Scotland Yard *dudes* warned of a terrorist attack. Our leader can be such a *klutz,* though, when explaining himself in public, that it sounded as if he didn't care *diddly* about his British counterpart. His problem is that he is *uptight* in front of the cameras and never quite learned how to *let it all hang out.*
>
> **REWRITTEN VERSION OF THE ABOVE:** The president decided to *cancel* his appointment with the British prime minister when the Scotland Yard *police* warned of a terrorist attack. Our leader can be *so awkward,* though, when explaining himself in public, that it sounded as if he didn't *have much respect for* his British counterpart. His problem is that he is *nervous* in front of the cameras and never quite learned how to *relax.*

split——————SPLIT INFINITIVE

Do not split infinitives unnecessarily. ***To speak, to go, to think*** **are infinitives. For example, you split the infinitive** ***to speak*** **when you place a word or words between** ***to*** **and** ***speak: to hastily speak*** **or** ***to now and then speak.*** **In formal writing, split infinitives are now acceptable if they read** ***smoothly.*** **Sometimes it is less awkward to split an infinitive than not to split it, but that is not often the case:**

> **UNACCEPTABLE:** He foolishly tried *to,* without studying at all, *pass* the chemistry final.
> **REVISED:** He foolishly tried *to pass* the chemistry final without studying at all.

> **UNACCEPTABLE:** He manages *to* usually *bore* people to death.
> **REVISED:** He usually manages *to bore* people to death.

ACCEPTABLE: He managed *to* completely *undermine* the work of the committee. [If you try to place *completely* elsewhere—before *to,* after *undermine,* or after *committee*—the sentence will not read as smoothly as it does *with* the split infinitive.]

SUBORDINATION —— sub

Emphasize important ideas by keeping them as main clauses. Change lesser ideas into subordinate clauses, phrases, and even single words where possible. **(See *Variety in Sentence Patterns.*)**

COORDINATE SENTENCE STRUCTURE

In sentence structure, *subordination* is the opposite of *coordination. Coordination* is the use of word groups that are structurally equal to express ideas that are equal in importance. Julius Caesar's "I came, I saw, I conquered" is a good example of main clauses arranged in a coordinate series. More typically, coordinate structures are joined by any of the following words, which are called coordinating conjunctions: *and, but, or, nor, for, yet, so.* For example, "I jog *and* I swim, *but* I do not play tennis."

SUBORDINATE SENTENCE STRUCTURE

Subordination is the use of word groups that are structurally unequal to express ideas that are unequal in importance: "I read the book because I liked the movie." Here the main idea is in the *main* clause, *I read the book,* and the less important, or subordinate, idea is in the *subordinate* clause *because I liked the movie.* The main clause, containing the main idea, can stand alone as a sentence: *I read the book*; but the subordinate clause, containing the less important idea, cannot.

Subordinate clauses begin with *subordinators*—subordinating conjunctions or relative pronouns. [*Subordinate clause* is defined under **Variety in Sentence Patterns.**] Here is a list of common subordinators:

Subordinating Conjunctions

after	because	provided	whenever
although	before	since	where
as	even though	so that	wherever
as if	if	though	while
as long as	in order that	unless	why
as soon as	no matter how	until	
as though	once	when	

Relative Pronouns

that	which	whoever	whomever
what	who	whom	whose

STRINGY SENTENCES: Do not use coordination—stringing ideas together with *and* or *so*—when subordination would better express the relationship of the ideas.

STRINGY: I saw the movie three times *and* I realized I always found it more fascinating each time, *so* I finally read the book.
BETTER: I finally read the book *because, after seeing* the movie three times, I realized I always found it more fascinating each time.

CHOPPY SENTENCES

Another form of abusing coordination is writing a series of short, choppy sentences even though the ideas are *not* of equal importance:

CHOPPY: People were bored. We became irritable. The picnic broke up early.

If we use *subordination,* however, we can show more clearly how these ideas are connected and express them in one smooth sentence:

BETTER: Becoming irritable out of sheer boredom, we broke up our picnic early.

(See **Choppy Sentences.**)

Further examples of sentences requiring the use of subordination follow:

STRINGY: John's employer did not care for him, *so* she refused to write him a letter of recommendation. [Two equally emphatic main clauses.]
BETTER: *Because* John's employer did not care for him, *she* refused to write him a letter of recommendation. [The first main clause, less emphatic than the second, is changed into a subordinate clause beginning with *because.*]

CHOPPY: She was exhausted. She had been swimming too long and was doubled up by a sudden cramp. She shouted for help. [This is an awkward series of choppy sentences.]
BETTER: Exhausted from swimming too long *and* doubled up by a sudden cramp, she shouted for help. [The first two sentences are turned into phrases.]

STRINGY: The moon *was glowing, and it* looked like the face of a snowman.
BETTER: The *glowing* moon looked like the face of a snowman. [The first main clause is condensed into the single word *glowing.*]

t ——— TENSE

1. **Check to see whether you are using the proper sequence of tenses. One of the verbs in your sentence may not be in the correct time relation with the other(s).**
2. **Use your dictionary to find the correct forms of irregular verbs (for example, *choose, chose, chosen*).**
3. **Do not shift tenses without good reason. (See *Shift in Point of View, 2.*)**
4. **Use the present tense to present plot summaries and your statements of an author's ideas.**

Tense is the form of a verb that tells your reader the *time*—past, present, or future—in which the action takes place. The verb *form* is the clue to the time. Here are three tense forms of the verb *work*: present (I *work,* he *works*); past (he *worked*); future (he *will work*).

1. SEQUENCE OF TENSES

If the time when an action takes place is the *same* in both the main clause and the subordinate clause, then the tense of both verbs must be the same:

- When she *arrived*, the crowd *greeted* her with a long ovation.
- As he slowly *turns,* he *balances* himself with his arms.

If the action in the subordinate clause takes place before that in the main clause, put the subordinate verb in the appropriate past tense:

- *I hear* that he *has worked* wonders.

The main verb, *hear,* is in the present tense; the subordinate verb, *has worked,* is in the present perfect tense. The present perfect tense expresses a time earlier than the present.

- *I heard* that he *had worked* wonders.

The past perfect, *had worked,* expresses a time prior to some understood time in the past. This *understood* past time is expressed by the simple past tense, *heard.*

When you are expressing a *permanent* fact, always use the present tense:

- I learned that the moon always *presents* the same face to the Earth. [Use *presents,* not *presented.*]

Keep an infinitive in the present tense if it expresses the same time as the action of the main verb; keep it in the past tense if it expresses a time before the action of the main verb:

- I would have liked *to go* with you.
- I would like *to go* with you.

In both these cases, although the main verb differs in tense, the present infinitive concerns *going* at the same time that the *liking* or the desire to go is expressed.

- I would like *to have gone* with you.

Here the past infinitive is used because the wish in the present concerns an action already completed in the past.

OVERKILL: I would *have liked* to *have gone* with you.

Do not use the past infinitive together with the past tense of the main verb. Use one or the other, as shown in the earlier examples, but not both at the same time.

2. IRREGULAR VERBS

Most English verbs are *regular,* forming their past tense and past participle with *-ed*: I *waited,* I have *waited.* With a regular verb like *wait,* once you know the present tense, you know all the other tenses. There is a troublesome group of *irregular* verbs, however, whose present tense (I *break*) is no clue to the past tense (I *broke*) or to the compound past tenses formed with the past participle (I have *broken*: *broken* is the past participle).

If you are in doubt about the past tense forms of a verb, look up the verb in the dictionary under its present-tense form (*bite,* for example) and you will find the past tense (*bit*) and past participle (*bitten*) listed in order right after it. Here is a list of some of the most frequently misused irregular verbs:

Present	Past	Past Participle
I *blow*	I *blew*	I have *blown*
I *break*	I *broke*	I have *broken*
I *bring*	I *brought*	I have *brought*
I *burst*	I *burst*	I have *burst*
I *buy*	I *bought*	I have *bought*
I *do*	I *did*	I have *done*
I *drink*	I *drank*	I have *drunk*
I *drive*	I *drove*	I have *driven*
I *eat*	I *ate*	I have *eaten*

Present	Past	Past Participle
I *find*	I *found*	I have *found*
I *fight*	I *fought*	I have *fought*
I *forbid*	I *forbade*	I have *forbidden*
I *go*	I *went*	I have *gone*
I *grow*	I *grew*	I have *grown*
I *lay* (bricks)	I *laid* (bricks)	I have *laid* (bricks)
I *lie* (down)	I *lay* (down)	I have *lain* (down)
I *leave*	I *left*	I have *left*
I *make*	I *made*	I have *made*
I *ring*	I *rang*	I have *rung*
I *rise*	I *rose*	I have *risen*
I *run*	I *ran*	I have *run*
I *see*	I *saw*	I have *seen*
I *seek*	I *sought*	I have *sought*
I *sing*	I *sang*	I have *sung*
I *steal*	I *stole*	I have *stolen*
I *swim*	I *swam*	I have *swum*
I *swing*	I *swung*	I have *swung*
I *take*	I *took*	I have *taken*
I *write*	I *wrote*	I have *written*

The most irregular verb of all, unluckily, is the one we use the most—the verb *to be*:

Present	Past	Past Participle
I *am*	I *was*	He/she/it *has been*
He/she/it *is*	He/she/it *was*	I/we/you/they *have been*
We/you/they *are*	We/you/they *were*	

3. TENSE SHIFTS

Changes in tense must occur for a good reason. In the following example, there is no justification for the shift:

> **SHIFT:** I *ran* to his house and *tried* to find him, but I *arrive* too late.
> **REVISED:** I ran to his house and tried to find him, but I *arrived* too late.

If it seems natural to you to use *arrive* rather than *arrived,* it may be that in your daily conversational habits you are not used to using, or even hearing, the past-tense endings of verbs in standard English. If this is so, ask your instructor to recommend materials that will help you practice the standard tense forms. (See also **-ed Error in -ed Endings.**)

4. PRESENT TENSE IN PLOT SUMMARIES

It does not matter whether the original story was written in the past tense, or whether the author whose ideas you are reporting is long dead. Always use the *present tense* under the following circumstances:

For plot summaries: In Part IV of Jonathan Swift's *Gulliver's Travels*, Gulliver *becomes* the guest of a race of highly civilized horses who *are* the masters of a slave-race of degenerate Yahoos. As Gulliver *discovers* to his horror, the physically and morally disgusting Yahoos *are* very similar to human beings.

For your statement of an author's ideas: In Gulliver's Travels, Swift *weighs* man against beast and *concludes* that so-called "civilized" man *is* a moral monster who *ranks* far below any beast.

trans —————— TRANSITIONS

Use a word or phrase to form a logical bridge, or *transition*, between two thoughts. The best transition to use is the one that most exactly expresses the logical relationship between two thoughts, sentences, or paragraphs.

Transitions are a special group of words and phrases that show how a piece of writing progresses logically from one idea to the next. Transitions connect parts of sentences, one sentence to another, and one paragraph to another. They express logical relations between ideas such as addition (*also, besides, furthermore*), contrast (*but, however, on the contrary*), result (*therefore, consequently*), and space or time (*beyond, in the distance, now, afterwards*). The following passage uses transitions of time (in italics):

> *In its earliest stages*, war consisted solely of battle for hunting grounds. *Afterwards*, war involved struggles for pasture. *Later*, war was fought for tilled or tillable land.

There are many ways of showing the logical linkage between ideas. Commonly used transitions follow:

Transitional Words

accordingly	eventually	later	second
actually	finally	likewise	similarly
afterward	first	meanwhile	soon
again	further	moreover	still
also	furthermore	nevertheless	then
and	gradually	next	therefore
before	hence	nonetheless	thereupon
beforehand	here	notwithstanding	this
besides	however	nor	too
but	indeed	now	
consequently	last	otherwise	

Transitional Phrases

after all	for this purpose	in sum
all in all	generally speaking	in the first place
all things considered	in addition	in the meantime
and yet	in any event	in the past
as a result	in brief	on the contrary
at length	in contrast	on the other hand
at the same time	in fact	on the whole
by the same token	in like manner	to be sure
for example	in other words	to sum up
for the most part	in short	to this end
for instance	in spite of (that)	

WEAK TRANSITION: She lost one fortune *and,* as if to spite fate, rapidly accumulated a second.
BETTER: She lost one fortune *but,* as if to spite fate, rapidly accumulated a second. [*But* more forcefully expresses the intended contrast.]

TRANSITION MISSING: I liked him. I thought his table manners needed improving. [The sudden contrast between these two thoughts is not smoothly bridged.]
BETTER: I liked him. *However,* I thought his table manners needed improving.

TRANSITION MISSING: On the whole, I think that educated people have made the best politicians. There are exceptions. [The second sentence follows too abruptly.]
BETTER: On the whole, I think that educated people have made the best politicians. *Of course,* there are exceptions.

Note the use of transitional words and phrases (italicized) in the following—somewhat shortened—paragraph by the philosopher Schopenhauer:

> What the address is to a letter, the title should be to a book; *in other words,* its main object should be to bring the book to those amongst the public who will take an interest in its contents. It should, *therefore,* be expressive. . . . The worst titles of all are those which have been stolen, *those, I mean,* which have already been borne by other books; for they are *in the first place* a plagiarism, and *secondly* the most convincing proof of a total lack of originality in the author. . . .

(For further information see **Coherence; Logic, 5;** and **Paragraph.**)

trite ———— TRITENESS

Rewrite the marked passage to eliminate the triteness. Trite writing is dull, commonplace, and uninteresting. The fault may lie in the thought or the

phrasing—and frequently in both. Use the following suggestions to eliminate triteness from your writing:

1. ***Use livelier verbs.*** Many common verbs do not make for specific, lively writing. Instead of writing, "I *had to eat* my sandwich quickly," write, "I *wolfed down* my sandwich." Which is more effective, "The elderly patient *walked slowly* down the hospital corridor," or "The elderly patient *shuffled* down the hospital corridor"?

2. ***Use precise, vivid adjectives.*** Many adjectives—*handsome, beautiful, nice, ugly*—are too vague to create a clear, specific picture in the reader's mind. You should communicate as precisely as possible the specific picture you have in mind. Instead of writing, "My father has a *handsome* face and *nice* eyes," write something like, "My father has a *weather-beaten sportsman's* face with *gentle brown* eyes."

3. ***Use effective figures of speech (comparisons).*** Sometimes comparing one thing to another does the job better than a written explanation containing many words. Instead of writing, "We tried to get him to confess, but he *would not tell us a thing,*" write, "We tried to get him to confess, but he *was as uncooperative as a rhinoceros.*"

CLICHÉS

A comparison that grows out of the situation you are writing about is likely to be fresh and appealing. A figure of speech that you have heard before is likely to be a cliché. Clichés are a special case of triteness. They are expressions that were once vivid and picturesque but are now so commonly used that they have lost their original force. Here are some examples of clichés (italicized):

- I got up *on the wrong side of the bed* this morning.
- He wanted to live out in *the wide open spaces.*
- They made progress *by leaps and bounds.*
- On picnics one can relax and enjoy *Mother Nature.*

- I felt *as cool as a cucumber.*
- No one suspected the *trials and tribulations* they went through.
- Her cousin was *as quiet as a mouse.*

vague ——————— VAGUENESS

Rewrite the marked section in clear, direct, precise language. Vague writing is often described as *foggy* or *cloudy* because it lacks substance. It relies heavily on generalization and lacks specific, concrete ideas and facts. (See *Abstract Expressions*; *Diction;* and *Logic,* 2.)

There is nothing wrong with a clear, substantial generalization; for example, "Students who do not read well are unable to write well." You may not agree with this statement, but at least you have something specific to argue over. Compare that statement with this: "Students with problems in some areas have other problems as well." This statement is fuzzy, insubstantial, and evasive. It does not challenge the reader to think about a clear issue. It makes a noise without making a point.

Vague writing is like a picture out of focus. To develop a clear, precise writing style, focus your mind on the idea or image you want to present before you commit it to paper. If you see it clearly in your mind's eye, you have a good chance of bringing it out clearly on paper. Foggy writing mirrors foggy thinking:

VAGUE: Professor Moss is a tough teacher whose personality turns me off.
CLEARER: Professor Moss grades much too harshly and is insulting to students who challenge his ideas.

VAGUE: I voted again for Mayor Dexter because her policies have helped the city improve in many ways, as we can see all around us.
CLEARER: I voted again for Mayor Dexter because she has erased the city's budget deficit, built a new hospital and library, and helped improve relations between the police and the public.

VARIETY IN SENTENCE PATTERNS——var

Develop a lively style by varying the structures and lengths of your sentences.

Good writers are always juggling a limited number of basic sentence patterns, balancing one against another to avoid monotony and to create a pleasing, rhythmic flow. You will find these basic structures easy to remember *because you know them already*; you already possess an array of skills that you are probably not aware you have. Substituting one pattern for another, we shall run a sample passage through a sequence of changes to show how virtually the same *ideas* can be expressed through a variety of *forms.*

I. STRUCTURAL VARIETY

Simple, Compound, and Complex Sentences

SAMPLE PASSAGE: We lost the first game. We vowed to even the score the next day. [Here we have two *simple sentences.* Each simple sentence contains only one subject– verb nucleus—"We lost," "we vowed." The sentences stand uninterestingly next to each other. Notice how in the following examples the use of certain standard word structures creates meaningful relationships between these now separate ideas.]
USING COORDINATION: We lost the first game, *but* we vowed to even the score the next day. [We now have a *compound sentence,* which is at least two simple sentences connected by a coordinating conjunction—*and, but, or, nor, for, yet,* or *so.* The connection by *but* ties these two separate thoughts into a relationship of contrast.]
USING A SUBORDINATE CLAUSE: *After we lost the first game,* we vowed to even the score the next day. [A *subordinate clause* consists of a subordinating conjunction— like *after, because, since, when, although*—followed by, at the least, a subject and its verb—*we lost.* Try substituting *although* for *after.*]
USING A RELATIVE CLAUSE: After we lost the first game, we vowed *that we would even the score the next day.* [*A relative clause* is a type of subordinate clause normally beginning with a relative pronoun such as *that, what, which, who,* or *whom.* Note: The combination of a main clause (simple sentence) with a subordinate clause results in a *complex sentence.* One of the ways to gain variety in sentence patterns is to create a pleasing alternation of *simple, compound,* and *complex* sentences.]

The following paragraph, from an essay by Robert Jay Lifton in *The Final Epidemic,* illustrates the skillful alternation of simple, compound, and complex sentences:

> Although the idea of apocalypse has been with us throughout the ages, it has been within a religious context—the idea that God will punish and even eliminate man for his sins. [complex] Now it is our own technology and we are doing it ourselves. [compound] Nor is

it only the nuclear threat. [simple] There are chemical warfare and germ warfare; destruction of the environment, the air we breathe or the ozone layer; and depletion of the world's resources, whether of energy or food. [simple]

Different Types of Phrases

SAMPLE PASSAGE: We lost the first game. We vowed to even the score the next day.

USING A PARTICIPIAL PHRASE: *Having lost the first* game, we vowed to even the score the next day. [A *participial phrase* is a group of words beginning with a participle, the *-ing* form of a verb: in our example, *having*. It acts as an adjective and modifies the subject, *we,* of the main clause it introduces.]

USING A GERUND PHRASE: *Losing the first game* made us vow to even the score the next day. [A *gerund phrase* looks like a participial phrase. It starts with a gerund, also the *-ing* form of a verb—in our example, *losing*—except that

a gerund or whole gerund phrase acts as a noun. Here it acts as the subject of a sentence whose verb is *made.*]
USING A PREPOSITIONAL PHRASE: *After that first-game defeat,* we vowed to even the score the next day. [A *prepositional phrase,* like *before work, inside the CIA,* or *after our defeat,* consists of a preposition followed by a noun—*defeat*—and any modifiers of that noun—*that first-game.*]

Here is a list of some common prepositions:

about	beside	from	on	until
above	between	in	over	up
after	but	into	since	with
around	by	like	through	within
at	during	near	to	without
before	except	of	toward	
below	for	off	under	

USING AN INFINITIVE PHRASE: *To have lost the first game* was such a blow that we vowed to even the score the next day. [An *infinitive phrase* starts with an infinitive—in our example, *to have lost*—which is followed by a noun, *game,* and any modifiers of that noun, *the first.* The infinitive phrase in this example acts as one whole noun, the subject of a sentence whose verb is *was.* Note that this sentence is also *complex,* consisting of a main clause beginning with *To have lost* and a subordinate clause beginning with *that.*]

For more information on sentence patterns, see **Subordination.** To learn how to knit sentences together to form a smooth paragraph, see **Transitions** and **Paragraph.**

Here is a brief paragraph, modified from J. E. Oliver's *Perspectives on Applied Physical Geography,* that achieves sentence variety by using subordination, coordination, and all of the phrase types we have discussed:

> *Making use of loud noises* [gerund phrase] has been tried all over the world as a means *to change the weather.* [infinitive phrase] *In Europe, for example,* [prepositional phrases] people have tried *to prevent hail-*

storms, [infinitive phrase] *for* [coordinating conjunction] hail has always caused considerable damage to vineyards. *To stop the hail from forming,* [infinitive phrase] farmers in northern Italy fired cannons at thunderclouds. Others felt *that they could stop storms* [relative clause] by *ringing church bells loudly.* [gerund phrase] Surprisingly, in some places *ringing bells* and *firing cannons* [gerund phrases] did seem to reduce the amount of crop damage by hail. This method became so popular *that it was finally outlawed.* [relative clause] Too many people were killed *by misfiring cannons* [prepositional phrase] and *by lightning* [prep. phr.] *striking bell towers* [participial phrase].

2. SENTENCE-LENGTH VARIETY

Good writers vary the pace and rhythm of their prose by mingling long, short, and medium-length sentences in any extended passage, as in the following paragraph (slightly modified from *Lunar Science: A Post-Apollo View,* by Stuart Ross Taylor):

> The *Apollo 11* landing on the moon took place on July 20, 1969, at 3:17:40 P.M., Eastern Standard Time, near the southern edge of Mare Tranquillitatis. [medium-length sentence] The site was named Tranquillity Base. [short] Astronauts Neil Armstrong and Edwin Aldrin collected 21.7 kilograms of samples in twenty minutes of hurried collecting toward the end of their two-hour sojourn (EVA, or extra vehicular activity) on the lunar surface. [medium to long] These samples were received in the quarantine facilities of the Lunar Receiving Laboratory in Houston on July 25. [short] Four weeks of intensive examination began. [short] A team of scientific workers (the Lunar Sample Preliminary Examination Team, or LSPET comprising eleven NASA scientists and fifteen other scientists from universities and government agencies) carried out preliminary geologic, geochemical, and biological examination of the samples, providing basic data for the Lunar Sample Analysis Planning Team (LSAPT). [long] Many of the first-order conclusions about the samples (such as their chemical uniqueness, their great age, and the absence of water, organic matter, and life) were established in this period. [medium]

Note the sequence of sentence lengths in Taylor's paragraph: medium/short/medium—long/short/short/long/medium.

wdy —— WORDINESS

Express your ideas in fewer words. Do not pad your sentences with unnecessary, repetitious phrasing. Avoid the unnecessary repetition of *words* and *ideas*:

WORDY: The novel *Don Quixote,* by Cervantes, is a novel that satirizes the dying age of chivalry. [Why repeat the word *novel*? A simple revision cuts out *four* needless words.]
BETTER: The novel *Don Quixote,* by Cervantes, satirizes the dying age of chivalry.

WORDY: *In my opinion,* I *personally believe* that our system of government is the best. [*In my opinion, personally,* and *I believe* are three ways of phrasing the same *idea.* Do not use them all at once.]
BETTER: I believe that our system of government is best.

WORDY: *In the modern world of today,* the human race is enjoying the fruits of a long technological revolution *that took place throughout the entire period of the machine age.*
BETTER: Today the human race is enjoying the fruits of a long technological revolution. [All that has been left out is repetition that adds nothing.]

(See **Repetition.**)

Where possible, use short, direct grammatical constructions:

INDIRECT: Bill made the salad, and *the cake was baked by Henrietta.* [Use the active voice instead of the passive. See **Passive Voice.**]
DIRECT: Bill made the salad, and Henrietta baked the cake.

TOO LONG: I was responsible for overall maintenance, but *it was* Phil *who* did most of the repair jobs.
SHORTER: I was responsible for overall maintenance, but Phil did most of the repair jobs.

Some common wordy expressions to avoid:

Wordy	Concise
Along the line of	About
At that point in time	At that point (*or* At that time)
Crisis situation	Crisis
Due to the fact that (he objected)	Because (he objected)
Emergency situation	Emergency
For a long period of time	For a long time
For the purpose of	For
In spite of the fact that (he left)	Despite (his leaving)
In the event that	If
The true facts	The facts

wp ——— WORD PROCESSING

Thanks to commonly available word-processing systems, the physical task of revision has never been easier. Keep in mind, however, a few simple cautions regarding the preparation and submission of word-processed, paper copy ("hardcopy," as it is called) to your instructor:

- ***"Saving" your work:*** In the course of writing directly into your computer or word processor, avoid accidental losses by frequently *saving* your work—manually, at least once every paragraph; or by setting your "automatic save" function for intervals of no more than about fifteen to twenty minutes apart. Little is more depressing than losing hours and hours of work because of mechanical or electrical system failure. It is also extremely important to save your document even further—onto a *diskette*—in case your hard drive should fail!
- ***Hardcopy:*** If you can, always print out hardcopy at the end of each writing session. Many of you know by experience that you cannot trust electronic systems completely. Finally, when you are ready to hand in your completed hardcopy document, always keep at least one back-up hardcopy version of the document for your own files.
- ***Formatting and pagination:*** Set your margins for between one and one and a half inches on all four sides, and always set your line-spacing to double-space, to provide ample room for your instructor's comments and revision symbols. Remember also to activate the page-numbering function of your system.
- ***Fonts:*** Most word-processing programs these days come with a great variety of fonts (styles of type). Avoid the temptation to use some weird or fancy font. Use something standard, like Courier or Times Roman, or the "default" font (the one your system employs if you don't choose a different one). Unusual fonts call unnecessary attention to themselves and can be hard on your instructor's eyes. Do not use a font that is entirely in *italics* (slanted letters) either. Reserve italics for emphasis (see **Italics**). Further, use a standard *size* font, preferably ten *cpi* (characters per inch) and ten to twelve points per character (*points* measure the size of

each individual letter). Finally, be sure that the print is dark enough for your instructor to read with ease.

- ***Spellcheckers, etc.:*** Automatic spellcheckers, dictionaries (thesauruses), and grammar checkers are sometimes helpful but totally mindless and often unreliable aids. Do not rely on them for the *major* work of correction and revision of your writing. Finally, *you* are solely responsible for all the problems eventually pointed out in your manuscript. (Computer spellcheckers are fine, but if your misspelled word happens to be another word—*your* for *you're*, or *it's* for *its*, for example—the spellchecker will not pick up the misspelling.

APPENDIX ON DOCUMENTATION STYLES — doc/app

1. CONSTRUCTING AN MLA STYLE "WORKS CITED" PAGE

Setting up the page: For an MLA style paper, begin the "Works Cited" section on a new page. Center the words "Works Cited" (without the quotation marks) on the first line. Below this heading, list your references in alphabetical order. Begin each reference on a new line at the right margin. If additional lines are needed for a given reference, indent them five spaces. Double space all text.

Reference Entries

General Guidelines

Titles:

- Capitalize the first word, last word, first word after a colon, and all major words of a title.
- Do not capitalize articles, prepositions, coordinating conjunctions, and the "to" in infinitives.
- Underline titles of books, journals, and magazines.
- Place in quotation marks: titles of short stories, short poems, short films, and episodes of radio/television series.
- Place in quotation marks: titles of articles in periodicals and of any other works that are part of a larger publication.

Names of Authors and Editors: List names exactly as they appear on the source and in the same order. If there are *more than three* authors (or editors), list only the name of the first author, followed by the abbreviation *et al.*

Books

List elements for each book entry in the following order:

1. Author(s). If the book is a collection, list the editor(s). If there are no authors or editors, begin with the title.
2. Title.
3. If applicable, the name of the translator.
4. If the book is a second or later edition, include the edition number or name (see example 3).
5. Publication information: list the first-mentioned city on the title page, the publisher's name, and date.

EXAMPLES

1. Book with a single author

Cornelius, Michael G. Creating Man. New York: Vineyard Press, 2000.

2. Book with two authors

Thomas, Dylan, and John Davenport. The Death of the King's Canary. New York: Viking, 1977.

3. Book with three authors

Baird, Russell N., Arthur T. Turnbull, and Duncan McDonald. The Graphics of Communication: Typography, Layout, Design, Production. Fifth Edition. New York: Holt, Reinhart and Winston, 1997.

4. Book with more than three authors

Foley, James D., et al. Computer Graphics: Principles and Practice. Second Edition. Reading, MA: Addison-Wesley Publishing Company, 1990.

5. Book with a translator or editor

Kundera, Milan. The Unbearable Lightness of Being. Trans. Michael Henry Heim. New York: Harper & Row, Publishers, 1987.

6. Edited collection or anthology

Kunz, Don, Ed. The Films of Oliver Stone. Lanham, MD: The Scarecrow Press, Inc., 1997.

Periodicals

List elements for each periodical entry in the following order:

1. Author(s).
2. Title of article.
3. Name of periodical.

4a. *For magazines,* the date and page numbers as in example 7. Newspapers follow the same format with the addition of the section number or letter before the page numbers.

4b. *For journals,* the volume number, date, and page numbers as in example 8. If a journal begins with page one in each issue, include the issue number as well as the volume number, as in example 9.

EXAMPLES

7. Magazine or newspaper article

Quinn, Helen R., and Michael S. Witherell. "The Asymmetry between Matter and Antimatter." Scientific American October 1998: 76-81.

8. Article in a journal that numbers pages continuously throughout a volume

Kamps, Ivo. "Historiography and Legitimization in Henry VIII." College English 58 (February 1996): 192-215.

9. Article in a journal that begins each issue with page number one

West, Alan, and Colin Martindale. "Creative Trends in the Content of Beatles Lyrics." Popular Music and Society 20.4 (Winter 1996): 103-125.

Work Reprinted in an Anthology or Collection

If a work is cited as a reprint in a collection, then the information about the work is cited first, followed by the designation "Rpt. in," followed by the citation for the collection. The page numbers of the work in the collection should appear last.

10. Work reprinted in an anthology

```
Shaw, Bernard. Mrs Warren's Profession. Rpt. in
    M. H. Abrams et al., Eds. The Norton
    Anthology of English Literature. Sixth
    Edition. Volume 2. New York: W. W. Norton
    and Company, 1993. pp. 1711-1754.
```

World Wide Web and Internet Sources

The general format for Internet sources in the MLA style is similar to that for printed resources. The main change is that additional information relating to the Internet is required. List elements for Internet entries in the following order:

1. Author(s), editor(s), or compiler(s).
2. Title of work.
3. Name(s) of editor(s) or translator(s).
4. Publication information for any print version of the source in the format used for that type of material.
5. Title of scholarly project, database, periodical, or professional or personal site. If the site has no title, then provide a description such as "Home page."
6. If available, name of the editor or compiler of the online scholarly project, website, or database.
7. Version number of the source, or for an online journal, the volume, issue, or other identifying number.
8. Date of electronic publication or posting, or the date of the latest revision.
9. If the source is a posting to a discussion group or electronic forum, include the name of the forum.
10. For articles: the number range or total number of pages, paragraphs, or sections.

11. Name of institution or organization sponsoring or associated with the website.
12. Date when YOU accessed the source as a reference.
13. Electronic address of the source in angle brackets:<> (NOTE: Electronic addresses are case-sensitive. Use the *exact* capitalization in the address.)

NOTE: In many cases, some of the above elements may not be found. Cite what is available. This may mean that the date of electronic publication and the date of access are consecutive.

EXAMPLES

11. A Scholarly Project or Reference Database (complete)

The William Blake Archive. Eds. Morris Eaves, Robert Essick, and Joseph Viscomi. 2 October 2001. 10 November 2001. <http://www.blakearchive.org/>.

12. A Document within a Scholarly Project or Reference Database

Blake, William. Visions of the Daughters of Albion. The William Blake Archive. Ed. Morris Eaves, Robert N. Essik, and Joseph Viscomi. 2 October 2001. 10 November 2001. <http://www.blakearchive.org/>.

"Mole Rat." Britannica Online. Vers. 98.2. April 1998. Encyclopaedia Britannica. 30 October 1998. <http://www.eb.com:180/>.

13. An Online Book

St. Augustine. Confessions. Trans., Ed. Albert C. Outler. [No Publisher Given], 1955. Ed. for WWW, Harry Plantinga and John Brubaker. 6 April 1994. The Wesley Center for Applied Theology. 30 October 1998. <http://www.iclnet.org/pub/resources/text/history/augustine/confessions.html>.

14. Article in a Journal

O'Neill, Robert V., James R. Kahn, and Clifford S. Russell. "Economics and Ecology: The Need for Détente in Conservation Ecology." Conservation Ecology 2.1 (June 1998): 4. 30 October 1998. <http://www.consecol.org/vol2/iss1/art4>.

15. Article in a Magazine

Brinkley, Douglas. "In the Kerouac Archive." The Atlantic Monthly November 1998: 49-76. 3 November 1998. <http://www.TheAtlantic.com/issues/98nov/kerouac.htm>.

16. Professional or Personal Site

Bolter, Jay David. Home page. 1998. <http://www.lcc.gatech.edu/~bolter/>. 25 November 1998.

Other Sources

EXAMPLES

17. Film

Total Recall. Dir. Paul Verhoeven. Perf. Arnold Schwarzenegger, Rachel Ticotin, Sharon Stone, Michael Ironside, and Ronny Cox. Carolco, 1990.

18. Film on Videocassette or Digital Video Disk

High and Low. Dir. Akira Kurosawa. Perf. Toshiro Mifune. 1963. DVD, Criterion Collection, 1998.

For other types of materials, consult the most recent edition of the *MLA Handbook for Writers of Research Papers.* The fifth edition was published in 1999.

2. CONSTRUCTING AN APA STYLE "REFERENCES" PAGE

Setting up the page: For an APA style paper, begin the "References" section on a new page. Center the word "References" (without quotation marks) on the first line. Below, list your references in alphabetical order. Begin each reference on a new line and indent it five spaces. If you need additional lines for a given reference, do not indent them. Double space all text on the page.

Reference Entries

General Guidelines

Titles of Books and Articles: Capitalize only the first word, the first word after a colon, and proper nouns. Underline titles of books. Do *not* use either underlining or quotation marks around article titles. Separate subtitles from titles with a colon.

Titles of Periodicals: Capitalize all major words of titles. Underline the titles of periodicals. For journals, continue the underlining through the volume number.

Names of Authors and Editors: List *every* author in the reference entry regardless of the number of names. Include only the author's last name and the first and middle initials even if the full names are given. For example, in a reference entry for the book *Toward an Aesthetic of Reception,* by Hans Robert Jauss, you would list the author's name as (Jauss, H. R.) as in example 4.

Books

List elements for each book entry in the following order:

1. Author(s). If the book is a collection, list the editor(s). If there are no authors or editors, list the title first.
2. Date of Publication in parentheses.
3. Title. If appropriate, the edition follows in parentheses as in example 1.
4. If necessary, the names of translator(s) or editor(s), as in examples 4 and 5.
5. Publication information. List the first city on the title page and the publisher.
6. If the book is a republication of a work, then the original publication date should be given as in example 5.

EXAMPLES

1. Book with a single author

Winston, P. H. (1984). Artificial intelligence
    (2nd ed.). Reading, MA: Addison-Wesley
    Publishing Company.

2. Book with two authors

Goodwin, F. K., & Jamison, K. R. (1990). Manic-depressive illness. New York: Oxford University Press.

3. Book with three or more authors

Press, W. H., Flannery, B. P., Teukolsky, S. A., & Vetterling, W. T. (1989). Numerical recipes in Pascal: The art of scientific computing. Cambridge, England: Cambridge University Press.

4. Book with a translator or editor

Jauss, H. R. (1982) Toward an aesthetic of reception (T. Bahti, Trans.). Minneapolis, MN: University of Minnesota Press.

5. Book that has been republished (this one has a translator who is also the editor)

Freud, S. (1977). Introductory lectures on psycho-analysis (J. Stratchey, Ed. and Trans.). New York: W. W. Norton and Company. (Original work published in 1917).

6. Edited collection or anthology

Mast, G., Cohen, M., & Braudy, L. (Eds.). (1992). Film theory and criticism:

Introductory readings (4th ed.). New York:
Oxford University Press.

Periodicals

List elements for each periodical entry in the following order:

1. Author(s). If there are no authors or editors, list the title first as in example 7.
2. Date of publication in parentheses.
3. Title of article (do not underline or place in quotation marks).
4. Name of periodical.

5a. For *newspapers,* skip to 6.

5b. For *magazines* and *journals* that number pages consecutively throughout a volume, list the volume number as in example 9. If a magazine or journal starts on page one in each issue, list the volume number followed by the issue number in parentheses, as in examples 8 and 10. The underlining of the title continues through the volume and issue number.

6. For all periodicals, page numbers for source. *For newspapers only,* include p. (single page) or pp. (multiple pages) immediately before the numbers. Example: "pp. A3–4." See example 7.

EXAMPLES

7. Newspaper article (no author given)

Normal EKGs could be deceiving, study shows.
(2000, October 24). The Providence Journal,
p. A3.

8. Magazine article

Quinn, H. R., & Witherell, M. S. (1998, October). The asymmetry between matter and antimatter. Scientific American, 279, 76-81.

9. Article in a journal that numbers pages continuously throughout a volume

Kamps, I. (1996, February) Historiography and legitimization in Henry VIII. College English, 58, 192-215.

10. Article in a journal that begins each issue with page number one

West, A., & Martindale, C. (1996, Winter). Creative trends in the content of Beatles lyrics. Popular Music and Society, 20(4), 103-125.

World Wide Web and Internet Sources

The format of the World Wide Web and Internet sources has changed in the fifth edition of the *Publication Manual of the American Psychological Association* (2001). A key element of this change is for sources that originally appeared in printed form but are available through electronic sources. If the electronic version of the source document is the same as the print version, then one should cite the document the same way that one would for the printed source with the notation [Electronic version] following the article or book title.

EXAMPLES

11. Journal article retrieved electronically that is the same as the printed text

Von Korff, M., & Goldberg, D. (2001, October). Improving outcomes in depression [Electronic version]. BMJ: British Medical Journal 323, 948-9.

If an electronic edition of a printed document shows changes (new materials, abridgement, revision, etc.), then give both the original publication information as well as information on the date and Internet location from which the article was retrieved. The following example gives the format:

12. Magazine article retrieved electronically that shows changes from the printed text

Brinkley, D. (1998, November). In the Kerouac archive. The Atlantic Monthly, 282. Retrieved November 3, 1998 from http://www.TheAtlantic.com/issues/98nov/kerouac.htm

List elements for Internet entries in the following order:

1. Author(s) or Editor(s). If absent, place title first. See example 7 for title first format.
2. Date of electronic publication or most recent revision. If unavailable, use the notation (n.d.) for "no date."
3. Title.
4. The name of the periodical if the source is an article.
5. If source is a periodical, the volume and issue number.

6. "Retrieved [exact date] from" followed by the electronic address of the source. Do not end with a period. (NOTE: Electronic addresses are case-sensitive. Use the *exact* capitalization in the address.)

EXAMPLES

13. Online book (with no online date provided on page)

James, W. (n.d.). The varieties of religious experience. (Original work published in 1902.) Retrieved October 3, 1998 from http://www.psychwww.com/psyrelig/james/toc.htm

14. Article in a reference database

Mole rat. (1998, April). In Britannica Online, Vers. 98.2. Retrieved October 30, 1998 from http://www.eb.com:180/cgi-bin/g?DocF=micro/399/57.html

15. Article in an online journal

O'Neill, R. V., Kahn, J. R., & Russell, C. S. (1998, June). Economics and ecology: The need for détente in conservation ecology. Conservation Ecology, 2.1, 4. Retrieved October 30, 1998 from http://www.consecol.org/vol2/iss1/art4

16. Personal or professional site

```
Bolter, J. (1998). [Home page]. Retrieved
    November 25, 1998 from
    http://www.lcc.gatech.edu/~bolter/
```

Other Sources

For other types of materials, consult the most recent edition of the *Publication Manual of the American Psychological Association.* The fifth edition was published in 2001.

INDEX

*Capitalized entries are section titles. Italicized entries are from "Words Often Misused: A Glossary" under the **Diction** section. Italicized page numbers are for main entries.